BIOENERGY ENGINEERING

The book provides information on recent advancements in bioenergy engineering to graduates, post-graduates, research scholars, faculty members, academician, researchers and practitioners studying and working in field of the bioenergy engineering. It is an invaluable information resource on biomass-based biofuels for fundamental and applied research, catering to researchers in the areas of biogas technology, densification techniques, biomass gasification, torrefaction of biomass, biochar production, micro algae production, improved biomass cookstoves, bio-ethanol production and the use of microbial processes in the conversion of biomass into biofuels. It will also be useful to faculties and researchers to understand the present status, advancements and policies in implementation of bioenergy technologies in India. This book will definitely provide a direction to the young researchers in identification of thrust areas of research in the field of bioenergy. The book concludes with research and development endeavours and aspects relating to implementation of advance bioenergy technologies.

Prof. Mahendra S. Seveda is Professor and Head, Department of Renewable Energy Engineering, Central Agricultural University, Ranipool, Gangtok, Sikkim, India since 2016. He is a Member of Executive Council, Sikkim University (Central University), Gangtok, Sikkim. He has worked as Associate Professor at Central Agricultural University, Ranipool, Gangtok, Sikkim, India from 2010 to 2015.

Dr. Pradip D. Narale, (born on March 04, 1989), is an Assistant Professor in Department of Renewable Energy Engineering, College of Agricultural Engineering and Post Harvest Technology, Central Agricultural University, Ranipool, Gangtok, Sikkim, India. After obtaining his B.Tech. degree in Agricultural Engineering from Vasantrao Naik Marathwada Agricultural University, Parbhani (Maharashtra), India in 2010; he obtained his M.E. and Ph.D. degree in Renewable Energy Engineering from Maharana Pratap University of Agriculture and Technology, Udaipur, Rajasthan, India in the year 2012 and 2017 respectively.

Dr. Sudhir N. Kharpude, (born on July, 1988), is an Assistant Professor in Department of Renewable Energy Engineering, College of Agricultural Engineering and Post Harvest Technology, Central Agricultural University, Ranipool, Gangtok, Sikkim, India. After obtaining his B.Tech. in Agricultural Engineering from Mahatma Phule Krishi Vidyapeeth, Rahuri, Maharashtra, India in 2010.

BIOENERGY ENGINEERING

Editors

Prof. Mahendra S. Seveda
Professor and Head
Department of Renewable Energy Engineering
College of Agricultural Engineering and Post Harvest Technology
Central Agricultural University, Ranipool, Gangtok-737135, Sikkim, India

Dr. Pradip D. Narale
Assistant Professor
Department of Renewable Energy Engineering
College of Agricultural Engineering and Post Harvest Technology
Central Agricultural University, Ranipool, Gangtok-737135, Sikkim, India

Dr. Sudhir N. Kharpude
Assistant Professor
Department of Renewable Energy Engineering
College of Agricultural Engineering and Post Harvest Technology
Central Agricultural University, Ranipool, Gangtok-737135, Sikkim, India

CRC Press is an imprint of the
Taylor & Francis Group, an **informa** business

NARENDRA PUBLISHING HOUSE
DELHI (INDIA)

First published 2022
by CRC Press
2 Park Square, Milton Park, Abingdon, Oxon, OX14 4RN

and by CRC Press
6000 Broken Sound Parkway NW, Suite 300, Boca Raton, FL 33487-2742

© 2022 selection and editorial matter, Narendra Publishing House; individual chapters, the contributors

CRC Press is an imprint of Informa UK Limited

The right of Mahendra S. Seveda, Pradip D. Narale and Sudhir N. Kharpude to be identified as the authors of the editorial material, and of the authors for their individual chapters, has been asserted in accordance with sections 77 and 78 of the Copyright, Designs and Patents Act 1988.

All rights reserved. No part of this book may be reprinted or reproduced or utilised in any form or by any electronic, mechanical, or other means, now known or hereafter invented, including photocopying and recording, or in any information storage or retrieval system, without permission in writing from the publishers.

For permission to photocopy or use material electronically from this work, access www.copyright.com or contact the Copyright Clearance Center, Inc. (CCC), 222 Rosewood Drive, Danvers, MA 01923, 978-750-8400. For works that are not available on CCC please contact mpkbookspermissions@tandf.co.uk

Trademark notice: Product or corporate names may be trademarks or registered trademarks, and are used only for identification and explanation without intent to infringe.

Print edition not for sale in South Asia (India, Sri Lanka, Nepal, Bangladesh, Pakistan or Bhutan).

British Library Cataloguing-in-Publication Data
A catalogue record for this book is available from the British Library

Library of Congress Cataloging-in-Publication Data
A catalog record has been requested

ISBN: 978-1-032-13799-5 (hbk)
ISBN: 978-1-003-23087-8 (ebk)

DOI: 10.1201/9781003230878

Contents

Preface .... *xvii*
List of Contributors .... *xix*

1. **Recent Trends and Future Prospects of Bioenergy Production in India** 1
   Anubhab Pal and Thaneswer Patel
   1. Introduction 2
   2. Biodiesel 3
      2.1 Present Status and Future Scope 3
      2.2 Potential Production and Consumption 5
   3. Biogas 6
      3.1 Present Status and Future Scope 6
      3.2 Potential Production and Consumption 8
   4. Bio-oil and Gasifier 10
      4.1 Present Status and Future Scope 10
      4.2 Potential Production and Consumption 11
   5. Sustainable Development 11
   6. Conclusions 14
      References 14

2. **Biogas Production, Utilization and Entrepreneurship Opportunities** .. 17
   Pradip D. Narale, Sudhir N. Kharpude and Mahendra S. Seveda 17
   1. Introduction 17
   2. Factors Affecting Biogas Production 19
      2.1 Waste Composition 19
      2.2 Volatile Solid 19
      2.3 Alkalinity and pH 19
      2.4 Volatile Fatty Acids Concentration 20

	2.5	Temperature	20
	2.6	Carbon to Nitrogen Ratio (C/N ratio)	20
	2.7	Hydraulic Retention Time (HRT)	21
	2.8	Organic Loading Rate (OLR)	21
3.	Status of Family Size Biogas Plants in India	21
4.	Classification of Biogas Plant	22
	4.1	Family Size Biogas Plants	22
		4.1.1 Fixed Dome Type Biogas Plant (Deenbandhu Biogas Plant)	22
		4.1.2 Floating Dome Type Biogas Plant (KVIC Biogas Plant)	23
		4.1.3 Prefabricated Biogas Plants	23
		4.1.4 Bag Type Biogas Plants (Flexi model)	25
	4.2	Industrial Biogas Plants	26
		4.2.1 Continuous Stirring Tank Reactor (CSTR)	26
		4.2.2 Up flow Anaerobic Sludge Blanket Reactor (UASB)	26
5.	Approved Models of Family Type Biogas Plants	27
6.	Hurdles in Implementing Family Size Biogas Plants in India	27
	6.1.	Availability of Feedstock	27
	6.2	Economic Constraints	28
	6.3	Promotion of Technology	28
	6.4	Availability of Implementing Agency or Skilled Labor at Root Level	28
	6.5	Social Dilemma	28
7.	Entrepreneurship Opportunities in Implementing Family Size Biogas Plants in India	28
8.	Application of Biogas	29
	8.1	Biogas used as a Cooking Fuel	29
	8.2	Use of Biogas as a Lighting Fuel	29
	8.3	Biogas for Power Generation	30
	8.4	Biogas used as Transportation Fuel	30
	8.5	Biogas used as a Heat Engine	31
	8.6	Cogeneration	31

		8.7	Utilization of Biogas for Water Pumping 31

        8.7   Utilization of Biogas for Water Pumping 31
        8.8   Biogas used in Fuel Cell Technology 31
        8.9   Biogas used in Refrigeration 32
   9.   Environmental Impacts and Sustainability 32
        9.1   Odour 32
        9.2   Pathogens 32
        9.3   Green House Gases 32
        9.4   Sustainability 32
        References 33

3. Advancements in Biogas Slurry Management Technologies 35
Madhuri More and Deepak Sharma

   1.   Introduction 36
   2.   Biogas Slurry Production 37
   3.   Applications of Biogas Slurry 39
        3.1   Organic Manure 39
        3.2   Composting Material 39
        3.3   Vermicomposting 39
        3.4   Bio-Pesticide Applications 39
        3.5   Others Applications 39
   4.   Utilization of Biogas Slurry 40
        4.1   Biogas Slurry Processing 40
        4.2   Advanced Technologies for Utilization of Biogas Slurry 40
            4.2.1 Screw Press Separation Technology 40
            4.2.2 Decanter Centrifuge Technology 42
            4.2.3 Belt Filter 43
        4.3   Use of Precipitation Agent for Separation and Settlement 44
            4.3.1 Processing of the Separated Solid Fraction 44
            4.3.2 Processing of the Separated Liquid Fraction 45
   5.   Economics of Biogas Slurry Processing 45
   6.   Conclusions 45
        References 46

4. **Technological Up-gradation in Biogas Production and Utilization for Energy Generation** 49
   Surendra R. Kalbande and Sejal R. Sedani
   1. Introduction 50
   2. Basic Process of Anaerobic Digestion 50
      2.1 Hydrolysis 50
      2.2 Acidogenesis (Acidification Phase) 51
      2.3 Acetogenesis 51
      2.4 Methanogenesis 52
   3. Limitations of Single-Stage Digesters 52
   4. Temperature Phased Anaerobic Digestion (AD) as an Alternative Technology to Improve Biomethanation 53
   5. Community Sized Biogas Plant for Electiricity Generation 54
      5.1 Main Components of Biogas Plant 54
   6. Biogas for Electricity Generation 56
   7. Performance of Biogas-Based Gas Engine 57
   8. Economic Analysis of the Biogas Power Generation System 58
   9. Evaluation of Performance of Community Sized 50 m3 Biogas Plant ... 59
      9.1 Characterization of Cattle Dung and Digested Slurry 59
   10. Analysis of Nutrients in Cow Dung and Digested Slurry 60
   11. Biogas Production 60
   12. Effect of Ambient Temperature on Biogas Production 61
   13. Composition of Biogas 61
   14. Performance of Gas Engine Runs on 100% Biogas 62
   15. Biogas-Enhancement Strategy 62
   16. Recent Advances in Biogas Purification Technologies 64
   17. Biogas Reforming Technologies 67
   18. Conclusions 68
   19. Acknowledgment 68
       References 68

5. **Present Status and Advancements in Biomass Gasification** 77
Hitesh Sanchavat, Vinit Modi, Tilak V. Chavda, Alok Singh and Sandip H. Sengar

1. Introduction 77
   1.1 Biomass Energy 77
   1.2 Biomass Energy and Its Sources 78
   1.3 Waste Types and Composition 79
   1.4 Biomass Properties 81
2. Biomass Energy Conversion Technologies 82
   2.1 Biomass Gasification 82
   2.2 Biomass Gasification Process 83
       2.2.1 Drying 83
       2.2.2 Biomass Pyrolysis 85
       2.2.3 Combustion 87
       2.2.4 Cracking 87
       2.2.5 Reduction 87
   2.3 Chemistry of Biomass Gasification 88
   2.4 Composition of Producer Gas 89
3. Classification of Gasifier 89
   3.1 Updraft Gasifier 90
   3.2 Downdraft Gasifier 91
   3.3 Fluidized Bed Gasifier 92
   3.4 Cross Draft Gasifier 92
   3.5 Twin Fire Gasifier 92
   3.6 Entrained-Flow Gasifier 92
   3.7 Other Types of Gasifiers 94
   3.8 Plasma Gasification for Toxic Organic Waste 95
   3.9 Supercritical Water Gasification (SCWG) for Wet Biomass 99
4. Sorption-Enhanced Reforming (SER) and Biomass Gasification with CO_2 Capture 100
5. Application of Syngas/Producer Gas 101
   5.1 Thermal Applications 101
   5.2 Power Applications 102

6. Advantages of Biomass Gasification Technologies 102
7. Cooling and Cleaning System of Gasifier 103
8. Factors Affecting Performance of the Gasifier 103
    8.1 Energy Content of the Feedstock 104
    8.2 Moisture Content 104
    8.3 Volatile Matter Content of The Fuel 104
    8.4 Particle Size and Distribution 104
    8.5 Bulk Density of Fuel 105
    8.6 Fuel Form 105
    8.7 Ash Content of Fuel 105
    8.8 Reactivity of Fuel 105
9. Important Terminologies 106
    References 106

6. Torrefaction of Biomass 113
Km. Sheetal Banga, Sunil Kumar and Raveena Kargwal
1. Introduction 113
2. What is Torrefaction? 114
3. Torrefaction Process Technique 115
    3.1 Reaction Temperature 115
        3.1.1 Core Temperature Rise 117
    3.2 Heating Rate 117
    3.3 Residence Time 117
    3.4 Biomass Type 117
    3.5 Ambience 120
4. Mechanism of Torrefaction 120
5. Torrefaction Products 121
6. Solid Torrefied Biomass Properties 122
    6.1 Physical Properties 122
        6.1.1 Moisture Content 122
        6.1.2 Bulk and Energy Density 122
        6.1.3 Grindability 122
        6.1.4 Pelletability 123

		6.2	Chemical Compositional Changes	123

      6.2 Chemical Compositional Changes 123
           6.2.1 Calorific Value 123
  7. Storage Aspects of Torrefied Biomass 123
      7.1 Off-gassing 123
      7.2 Hydrophobicity 124
  8. Applications with Torrefied Biomass 124
  9. Classification of Reactors Used in Torrefaction 125
10. Advantages of Torrefied Biomass 127
11. Conclusions 127
      References 129

7. Algal Biomass: A Promising Source for Future Bioenergy Production 131
Subodh Kumar, Adya Isha, Ram Chandra, Anushree Malik and Virendra K. Vijay
  1. Introduction 132
  2. Classification of Algal Biomass 133
      2.1 Characteristics of Macroalgae 133
      2.2 Characteristics of Microalgae 133
  3. Cultivation of Algal Biomass 133
      3.1 Cultivation of Microalgae 133
      3.2 Cultivation of Macroalgae 135
  4. Biogas Production from Algal Biomass 136
      4.1 Anaerobic Digestion Process 136
      4.2 Factor Affecting Anaerobic Digestion Process 137
      4.3 Biogas Production from Microalgae and Macroalgae 137
  5. Bioethanol Production from Algal Biomass 139
      5.1 Bioethanol Fermentation Process 140
      5.2 Bioethanol Potential of Microalgae and Macroalgae 142
  6. Biodiesel Production from Algal Biomass 142
      6.1 Biodiesel Production Process 143
           6.1.1. Pretreatment 143
           6.1.2 Fractionation 143
           6.1.3 Transesterification 144
  7. Conclusions 144
      References 145

8. Micro Algae Production for Bio Fuel Generation 153
Swapnaja K. Jadhav, Anil K. Dubey, Mayuri Gupta,
Sachin Gajendra and Panna Lal Singh

   1. Introduction 153
   2. What are Micro Algae? 154
   3. Factors Influencing Microalgae Growth 156
       3.1   Physical Factor 156
             3.1.1 Light Intensity 156
             3.1.2 pH and Salinity 156
             3.1.3 Nitrogen/Phosphorus Nutrient 156
             3.1.4 Temperature 156
             3.1.5 Carbon Dioxide 157
             3.1.6 Oxygen 157
             3.1.7 Water Requirement 157
       3.2   Biotic Factors 157
             3.2.1 Invasive Species and Predators 157
   4. Production of Micro Algae 158
       4.1   Up-Scaling of Micro-Algae Production 158
       4.2   Microalgae Cultivation 159
             4.2.1 Open Ponds 159
             4.2.2 Raceway Ponds 159
             4.2.3 Thin Layer Cascades 160
             4.2.4 Circular Ponds 160
             4.2.5 Closed systems/ Photo bioreactors (PBR) 160
             4.2.6 Hybrid Systems 161
       4.3   Comparison of Cultivation Systems 161
       4.4   Advantages and Disadvantages of Micro Algae
           Production Systems 162
             4.4.1 Advantages and Disadvantages of Open Pond System 162
             4.4.2 Advantages and Disadvantages of Closed system 162
   5. Construction of Open Pond Systems for Micro Algae Production 163
       5.1   Operating Factors to be Consider While Development
           of Open Ponds 163

	5.1.1 Pond Liners 163
	5.1.2 Mixing 163
	5.1.3 Depth 163
	5.1.4 Cell Concentration 164

    5.2 Construction Design of Raceway and Circular Types Open Ponds 164

6. Harvesting of Microalgae 166
7. Drying of Harvested Algal Biomass 168
8. Biofuels Derived from Microalgae 168
9. Status, Challenges and the Way Forward 168
    References 169

9. Biochar Production for Environmental Application 173
Ashish Pawar and Narayan L. Panwar

1. Introduction 173
2. Biochar 174
3. Status in Indian Context 175
4. Biochar Production Technologies 175
    4.1 Batch Type 176
    4.2 Continuous Type 176
    4.3 Novel Process 179
5. Factors Affecting Biochar Production 179
6. Physicochemical Properties of Biochar 180
7. Applications of Biochar 181
    7.1 Soil Conditioner 183
    7.2 Waste Water Treatment 184
    7.3 Energy Production 184
    7.4 Cosmetics 185
    7.5 Paints and Colouring 185
8. Advantages and Disadvantages of Biochar 185
    8.1 Advantages of Biochar 185
    8.2 Disadvantages of Biochar 186
9. Economics of Biochar 187
10. Conclusions 187
    References 187

10. Advancement in Improved Biomass Cookstove and Its Current Status in India 193
Himanshu Kumar, Amit Ranjan Verma, Swapna S. Sahoo and Narayan L. Panwar 193
1. Introduction 194
2. Classification of Biomass Cookstove 195
3. Recent Advancement in Cookstove Development 196
4. Potential to Mitigate GHG Emissions Through ICS 201
5. Policy Framework for Dissemination of Improved Cookstove in India 202
6. Problems of Lower Acceptance of Improved Biomass Stove 203
7. Strategy to Enhance Adoptability of ICS– Author's View 204
8. Conclusions 204
   References 205

11. Practical Evaluation Approach of a Typical Biomass Cookstove 209
Amit Ranjan Verma, Ratnesh Tiwari, Manoj Kumar Verma and Himanshu Kumar 209
1. Introduction 210
2. Need for Cookstove Testing 210
   2.1 Types of Testing 211
      2.1.1 Laboratory Testing 211
      2.1.2 Field Testing 211
   2.2 Performance Parameters 212
      2.2.1 Thermal Performance Parameters 212
         2.2.1.1 Power 212
         2.2.1.2 Thermal Efficiency and Specific Fuel Consumption 213
         2.2.1.3 Turn-Down Ratio 213
      2.2.2 Emission Performance Parameters 213
      2.2.3 Effect of Operating Parameters 213
         2.2.3.1 Fuel Type 213
         2.2.3.2 Fuel Size 214
         2.2.3.3 Fuel Moisture Content 214
         2.2.3.4 Pot Size and Lid 214

3.	Stove Testing Protocols	214
	3.1 History and Evolution of Testing Protocols	214
	3.2 Performance Evaluation of Biomass Cookstoves	216
	3.3 Uncertainty in Testing	224
4.	Conclusions	224
	References	226

12. Densification Technologies for Agro Waste Management 231
Abolee Jagtap and Surendra R. Kalbande

1.	Introduction	231
2.	Mechanics of Bonding of Particles	233
3.	Raw Materials for Briquetting and Pre-treatments	234
	3.1 Collection of Materials	234
	3.2 Pre-treatment of Raw Materials	234
4.	Densification Technologies	236
	4.1 Piston Press Type Machine	236
	4.2 Screw Press Type Machine	237
	4.3 Roller Press Type	237
	4.4 Manual Press and Low Press Machine	238
	4.5 Flat Die Type Machine	239
5.	Comparison of Different Densification Technologies	239
6.	Characteristics of Biomass Briquettes	241
	6.1 Physical Properties	241
	6.1.1 Moisture Content	241
	6.1.2 Bulk Density	241
	6.1.3 Shatter Resistance Test	242
	6.1.4 Tumbling Resistance Test	242
	6.1.5 Water Penetration Resistance	243
	6.1.6 Degree of Densification	243
	6.1.7 Energy Density Ratio	243
	6.2 Thermal Properties of Briquettes	243
	6.2.1 Volatile Matter	244
	6.2.2 Ash Content	244

		6.2.3 Fixed Carbon 244

- 6.2.3 Fixed Carbon 244
- 6.2.4 Calorific Value 245
- 6.3 Energy Evaluation Analysis 245
  - 6.3.1 Thermal Fuel Efficiency (TFE) Test 245
  - 6.3.2 Burning Rate 246
  - 6.3.3 Ignition Time 246
- 7. Economics of Briquetting Technology 247
- 8. Binders Used in Biomass Densification Technology 248
  - 8.1 Lignosulfonates 248
  - 8.2 Bentonite 249
  - 8.3 Starches 249
  - 8.4 Protein 249
- 9. Conclusions 249
- References 250

13. Recent Advancement in Biochemical Conversion of Lignocellulosic Biomass to Bioethanol and Biogas 253
Sweety Kaur, Richa Arora and Sachin Kumar 253

1. Introduction 254
2. Microorganisms Involved in Biofuel Production 256
3. Substrates Utilized by Microorganisms for Production of Biofuels 258
4. Biosynthesis of Bioethanol and Biogas 260
5. Second Generation Biofuels 262
6. Related Challenges and Possible Solutions 263
   References 264

14. Bamboo as a Building Material for Climate Change Mitigation 269
Vishal Puri

1. Introduction 270
2. Bamboo in Construction 270
   - 2.1 Bamboo as Reinforcement in Structural Elements 272
   - 2.2 Bamboo for Climate Change Mitigation 272
3. Codal Provisions 273
4. Conclusions 273
   References 276
   Index 279

Preface

This book, entitled *Bioenergy Engineering,* is designed keeping in view the bioenergy engineering course curricula prescribed by IITs, NITs, engineering colleges, central universities, state universities, technical universities of India and abroad.

The book consolidates the most recent research on current technologies, concepts and commercial developments in various types of widely used biofuels.

The 14 chapters of this book offer sate of the art reviews, concepts and methodologies current research, and technology developments with respect to biomass energy conversion technologies and its application for various energy generation and utilization prospects. Chapters are also designed to facilitate early stage researchers and enable to easily grasp the concepts, methodologies and application of bioenergy technologies. It is a complete book in all respect of bioenergy.

It includes basics and recent trends of bioenergy production in India, production and utilization of biogas, advancements in biogas slurry management, technological upgradation of biogas production and utilization, biomass densification techniques, biomass gasification, torrefaction of biomass, biochar production, improved biomass cookstoves, practical approach in evaluation of biomass cookstoves, algal biomass, biochemical conversion of biomass into bio-ethanol and biogas and bamboo as building material for climate change. Each topic has been discussed in detail, both conceptually and methodologically, so that students do not face any kind of difficulties.

The scope of the book thus has been expanded beyond the basic needs of undergraduate and post graduate engineering students. We hope this book will be of immense help not only to the students but also to the faculty members, researchers, academicians, scientists, teachers, policy makers, entrepreneurs, extension workers and professionals.

We are very grateful to all the authors for their contributions and the reviewers their book chapters, thus improving the quality of this book.

Lastly, we would like to express our thanks and sincere regards to our family members who have provided great support to us.

Mahendra S. Seveda
Pradip D. Narale
Sudhir N. Kharpude

List of Contributors

Dr. Anubhab Pal
Assistant Professor,
Department of Agricultural Engineering,
North Eastern Regional Institute of
Science and Technology, Nirjuli,
Itanagar, Arunachal Pradesh, India

Dr. Thaneswer Patel
Assistant Professor,
Department of Agricultural Engineering,
North Eastern Regional Institute of
Science and Technology,
Nirjuli, Itanagar, Arunachal Pradesh, India

Dr. Pradip D. Narale
Assistant Professor,
Department of Renewable Energy
Engineering,
College of Agricultural Engineering and
Post Harvest Technology,
Central Agricultural University,
Ranipool, Gangtok, Sikkim, India

Dr. Sudhir N. Kharpude
Assistant Professor,
Department of Renewable Energy
Engineering,
College of Agricultural Engineering and
Post Harvest Technology,
Central Agricultural University,
Ranipool, Gangtok, Sikkim, India

Prof. Mahendra S. Seveda
Professor and Head,
Department of Renewable Energy
Engineering,
College of Agricultural Engineering and
Post Harvest Technology,
Central Agricultural University,
Ranipool, Gangtok, Sikkim, India

Er. Madhuri More
Research Scholar,
Department of Renewable Energy
Engineering,
College of Technology and Engineering,
Maharana Pratap University of
Agriculture and Technology,
Udaipur, Rajasthan, India

Prof. Deepak Sharma
Former Head,
Department of Renewable Energy
Engineering,
College of Technology and Engineering,
Maharana Pratap University of
Agriculture and Technology,
Udaipur, Rajasthan, India

Prof. Surendra R. Kalbande
Professor and Registrar,
Dr. Panjabrao Deshmukh Krishi
Vidyapeeth, Krishi Nagar,
Akola, Maharashtra, India

Er. Sejal R. Sedani
Senior Research Fellow,
Department of Unconventional Energy
Sources and Electrical Engineering,
Dr. Panjabrao Deshmukh Krishi
Vidyapeeth, Krishi Nagar,
Akola, Maharashtra, India

Er. Abolee Jagtap
Senior Research Fellow,
Department of Unconventional Energy
Sources and Electrical Engineering,
Dr. Panjabrao Deshmukh Krishi
Vidyapeeth, Krishi Nagar,
Akola, Maharashtra, India

Dr. Km. Sheetal Banga
Assistant Professor,
Janta Vedic College, Baraut,
Chaudhary Charan Singh University
Meerut, Uttar Pradesh, India

Er. Tilak V. Chavada
Assistant Professor,
Department of Renewable Energy
Engineering,
College of Agricultural Engineering and
Technology,
Navsari Agricultural University,
Dediapada, Gujarat, India

Er. Vinit Modi
Research Scholar,
Center of Excellence in Energy Studies
and Environment Management,
Deenbandhu Chhotu Ram University of
Science and Technology, Murthal,
Sonepat, Haryana, India

Er. Alok Singh
Assistant Professor,
Department of Renewable Energy
Engineering, College of Agricultural
Engineering and Technology,
Navsari Agricultural University,
Dediapada, Gujarat, India

Dr. Sandip H. Sengar
Associate Professor and Head,
Department of Renewable Energy
Engineering, College of Agricultural
Engineering and Technology,
Navsari Agricultural University,
Dediapada, Gujarat, India

Er. Raveena Kargwal
Research Scholar,
College of Agricultural Engineering and
Technology, Chaudhary Charan Singh
Haryana Agricultural University,
Hisar, Haryana, India

Er. Ashish Pawar
Research Scholar,
Department of Renewable Energy
Engineering,
College of Technology and Engineering,
Maharana Pratap University of
Agriculture and Technology,
Udaipur, Rajasthan, India

Dr. Narayan L. Panwar
Assistant Professor,
Department of Renewable Energy
Engineering,
College of Technology and Engineering,
Maharana Pratap University of
Agriculture and Technology,
Udaipur, Rajasthan, India

Er. Swapna S. Sahoo
Research Scholar,
Centre for Rural Development and
Technology,
Indian Institute of Technology Delhi,
Hauz Khas, New Delhi, India

Er. Himanshu Kumar
Research Scholar,
Centre for Rural Development and
Technology,
Indian Institute of Technology Delhi,
Hauz Khas, New Delhi, India

Er. Amit Ranjan Verma
Research Scholar, Centre for Rural
Development and Technology,
Indian Institute of Technology Delhi,
Hauz Khas, New Delhi, India

Dr. Ratnesh Tiwari
Centre for Rural Development and
Technology, Indian Institute of
Technology Delhi, Hauz Khas,
New Delhi, India

Er. Manoj Kumar Verma
Research Associate,
Centre for Rural Development and Technology, Indian Institute of Technology Delhi, Hauz Khas, New Delhi, India

Er. Subodh Kumar
Research Scholar,
Centre for Rural Development and Technology, Indian Institute of Technology Delhi, Hauz Khas, New Delhi, India

Er. Adya Isha
Research Scholar,
Centre for Rural Development and Technology,
Indian Institute of Technology Delhi, Hauz Khas, New Delhi, India

Dr. Ram Chandra
Assistant Professor,
Centre for Rural Development and Technology, Indian Institute of Technology Delhi, Hauz Khas, New Delhi, India

Prof. Anushree Malik
Professor and Head, Centre for Rural Development and Technology, Indian Institute of Technology Delhi, Hauz Khas, New Delhi, India

Prof. Virendra Kumar Vijay
Professor, Centre for Rural Development and Technology, Indian Institute of Technology Delhi, Hauz Khas, New Delhi, India

Er. Swapnaja Jadhav
Scientist,
Agricultural Energy and Power Division, Central Institute of Agricultural Engineering, Navi Bagh, Bhopal, Madhya Pradesh, India

Dr. Anil K. Dubey
Former Principal Scientist
Agricultural Energy and Power Division, Central Institute of Agricultural Engineering,
Navi Bagh, Bhopal, Madhya Pradesh, India

Er. Sunil Kumar
Assistant Scientist,
Department of Processing and Food Engineering,
College of Agricultural Engineering and Technology,
Chaudhary Charan Singh Haryana Agricultural University,
Hisar, Haryana, India

Dr. Hitesh Sanchavat
Assistant Professor and Head,
Department of Farm Machinery and Power Engineering,
College of Agricultural Engineering and Technology,
Navsari Agricultural University,
Dediapada, Gujarat, India

Dr. Mayuri Gupta
Senior Research Fellow,
Consortia Research Platform on Energy in Agriculture,
Agricultural Energy and Power Division, Central Institute of Agricultural Engineering,
Navi Bagh, Bhopal, Madhya Pradesh, India

Er. Sachin Gajendra
Senior Research Fellow,
Consortia Research Platform on Energy in Agriculture,
Agricultural Energy and Power Division, Central Institute of Agricultural Engineering, Navi Bagh, Bhopal, Madhya Pradesh, India

Dr. Panna Lal Singh
Principal Scientist,
Agricultural Energy and Power Division,
Central Institute of Agricultural Engineering,
Navi Bagh, Bhopal,
Madhya Pradesh, India

Dr. Richa Arora
Microbiologist,
Department of Microbiology,
College of Basic Sciences and Humanities,
Punjab Agricultural University,
Ludhiana, Punjab, India

Dr. Vishal Puri
Assistant Professor,
Department of Civil Engineering,
J.C Bose University of Science and Technology YMCA,
Faridabad, Haryana, India

Dr. Sachin Kumar
Deputy Director/Scientist,
Biochemical Conversion Division,
Sardar Swaran Singh National Institute of Bio-Energy,
Kapurthala, Punjab, India

Ms. Sweety Kaur
Nestle India Limited,
Rajarhat, Kolkata, India

CHAPTER - 1

RECENT TRENDS AND FUTURE PROSPECTS OF BIOENERGY PRODUCTION IN INDIA

Anubhab Pal and Thaneswer Patel*

Department of Agricultural Engineering
North Eastern Regional Institute of Science and Technology
Nirjuli, Itanagar- 791109, Arunachal Pradesh, India
*Corresponding Author

Indian economy is based on agriculture, and therefore, biomass has been an important energy source in Indiafor a long time. Currently, India has an energy potential of about 18 GW of energy from biomass. About 70% of the country's population depends upon biomassfor theirdaily energy needs. The energy demand in India is expected to be more than 3 to 4 times the current level in the next 25 years. Hence Ministry of New and Renewable Energy has initiated several schemes and programme for promotingand popularizingof efficient bioenergy production technologies for its use to ensure the derivation of maximum benefits. Bioenergyis being used for various purposes such as heat generation, transport, electricity generation etc. Currently, the world is facing a major climate change crisis due to greenhouse gas emissions from petroleum products, and therefore in many countries,including India, over the past few years, many researchers have actively been investigating the adaptability of renewable energy sources to different applications. As s result, the contribution of bio-energy to heat in industries has grown by around 2% during recent years. Bioenergy can be classified into two types based on the conversion technique used viz. thermochemical and biochemical conversion. The bio-energy produced by the thermochemical process is mainly used in domestic and as well as industrial purposes and the bio-energy produced by biochemical sources is commonly used in transportation. In this chapter, the status, potential, limitations, and future research direction of bioenergy production have been discussed.

1. INTRODUCTION

Bioenergy is the energy derived from the organic material, also known as biomass. The energy may be in the form of gas – biogas, liquid – biodiesel etc. The biomass can be from plant timber, leaves, agricultural waste, animal waste, and sometimes sewage. The biomass used as an energy source is known as a feedstock. The feedstocks are sometimes grown especially for their energy content, or sometimes these feedstocks can be made up of the waste products obtained from agriculture and allied industries. In India, for a long time, biomass is being used to meet the energy demands of the rural population. Biomass in the form of firewood and cowdung cake has always been used by our forefathers for cooking and heating. 21.67 % (712,249 ha) of India's total geographical area is covered by forest, and hence there is a massive scope of biomass harvesting from these forest land. As biomass is renewable, available widely,and carbon-neutral, therefore it has a lot of potential for providing a significant level of employment in ruralIndia through small and cottage industries. Another benefit of biomass is that it is also capable of providing energy securitytorural India.

According to the Ministry of New and Renewable Energy, Government of India, New Delhi around 32% of the total primary energy being used in India is derived from biomass. Also, more than 70% of the country's population depends upon the biomass for their energy needs [1]. As India is a signatory of the Paris climate agreement, therefore, by the year 2030, India has to create a cumulative carbon sink of 2.5-3 billion tonnes of CO_2 equivalent. This can be achieved by reducing the dependency on fossil fuel and provide more focus on renewable sources like bioenergy. India's potential for bioenergy is around 18 GW and has ~5+ GW capacity biomass power plants, among which 83% are grid-connected, and the remaining 17% are off-grid plants. Also, around 70 Cogeneration projects are being implemented shortly with an additional capacity of 800 MW [2]. As a result of the instability in the oil prices in the International market and also because of the instability in the oil-producing regions, the surge in the energy demand from both rural and as well as urban areas, and also due to the greater awareness about the climate change and its implications due to fossil fuel usage have positively contributed to a greater interest in bioenergy among the government and as well as the common citizens.In the following chapter, we will discuss the different sources of bioenergy, their scope, and future prospects in the Indian context.

2. BIODIESEL

India is the world's third-largest oil consumer after the USA and China. It is expected that the rate of India's oil consumption will surpass China's oil consumption in the near future. The major contributor to this huge demand in oil is electricity generation and the transportation sector. In the financial year, 2019-20 India consumed about 29,976 thousand metric tonnes of petrol (14.0%), 82,579 thousand metric tonnes of diesel (38.6%) out of total 2,13,686 thousand metric tonnes of petroleum consumption [3]. It is clear from the data that most of the petroleum being used by India is in the form of diesel, and hence it can be easily considered as one of the highest contributors of greenhouse gases in India. Therefore there is a need for the shift into the from the conventionally used diesel derived from petroleum to a source that is more sustainable and also carbon neutral. The consumption of different petroleum products of India is shown in Table 1.1.

Biodiesel is a type of renewable fuel produced by chemically reacting to any natural oil or fat with alcohol, viz. methanol or ethanol. It is a sustainable source of energy and can be actively used as a replacement of petrodiesel in both transportation and energy generation sectors. Biodiesel can be produced from any natural oils which can be produced domestically and hence it can help in reducing India's dependency on imported petroleum products. Also, the vast unused lands can be used for energy plantation to which can supply oilseeds to harvest vegetable oil for biodiesel. It will also help in employment generation in rural India. These domestically grown vegetable oils will offer new energy-related markets and help reduce the import of petroleum products and thus reduce international spending and improve the economy. Biodiesel can be used neat (100%) or as a blend with the petro-diesel. It has been found that the emissions from vehicles running on biodiesel vary linearly with the blend level. One litreof biodiesel provides similar benefits when used at 100% concentration, or used in blends, with petrodieselsuch as B20 (20% biodiesel with 80% diesel fuel) [4].

2.1 Present Status and Future Scope

For the financial year 2019-20, out of total demand of 2,13,686 thousand metric tonnes of petroleum products, India imported 2,70,285 thousand metric tonnes (~80%)[5].The diesel share is around 38.6% of the total petroleum productconsumption.The requirement of Diesel in India is over 2.75 times higher than petrol, and the consumption of diesel has been increased from 56,242 thousand tonnes in FY 2009-10 to 82,579 thousand tonnes in FY 2019-20[5]. That means an increase in 1.5 times over a decade. Furthermore, thenumber of vehicles

Table 1.1: Petroleum product Consumption in India for 2019-20 FY[3]

Products	Consumption of Petroleum Products in India in 1000 Metric onnes												
	AP	MA	JU	JU	AU	SE	OC	NO	DE	JA	FE	MA	Total
Liquidied Petroleum Gas	1,900	2,054	1,793	2,219	2,402	2,164	2,348	2,261	2,354	2,449	2,115	2,306	26,366
Naphtha	949	795	987	1,465	1,420	1,107	1,132	1,257	1,277	1,383	1,279	1,386	14,436
Petrol	2,459	2,737	2,639	2,523	2,575	2,372	2,539	2,535	2,473	2,456	2,511	2,156	29,976
Aviation Turbine Fuel	645	679	650	658	666	653	697	709	729	740	690	484	8,000
Superior Kerosene Oil	254	268	260	195	231	176	170	188	153	164	185	152	2,397
High Speed Diesel	7,323	7,788	7,451	6,841	6,117	5,837	6,510	7,571	7,387	6,942	7,160	5,651	82,579
Light Diesel Oil	45	49	53	52	63	60	48	51	46	57	54	49	628
Lubricants & Greases	255	354	300	305	311	310	277	292	284	327	326	296	3,640
Furnace Oil & Low Sulphur Heavy Stock	499	515	476	568	479	529	472	460	626	486	503	482	6,094
Bitumen	691	726	548	407	231	321	446	585	629	598	670	525	6,377
Petroleum coke	2,254	2,135	1,480	1,495	1,783	1,757	1,667	1,676	2,002	1,946	1,786	1,680	21,659
Others	1,045	982	1,037	1,020	903	897	1,000	896	906	986	946	917	11,535
Total	18,320	19,083	17,674	17,749	17,181	16,185	17,308	18,479	18,866	18,535	18,223	16,083	2,13,686

running on diesel has also been increased over the years. As of 2015, the total number of registered buses, passenger LMV and Goods vehicle was 1,59,00,172[6], whereas, in the year 2019, the number was 1,03,93,336. That means an increase of around 53% over merely seven years.

In India, it is impractical to produce biodiesel from vegetable oils obtained from edible oilseeds as India is heavily dependent upon them for food security.Therefore, non-edible oil sources such as Jatropha, Karanja etc. are being considered as alternative feedstocks for biodiesel production [7]. National Biofuels Policy was established in 2009 to provide energy security and meet the higher demand forfuel in the upcoming years. The objectives of thepolicy wereutilization of non-food feedstock and to improve research on their cultivation in the wastelandsand also bringing the blending mandate of 20% by 2017 [8]. Several state and national level organizations and as well as private organizations such as Gujarat Oelo Chem Limited, Reliance Industries Ltd, Aatmiya Biofuels Pvt Ltd., Godrej Agrovet, Nova Bio Fuels Pvt. Ltd., Jain Irrigation System Ltd., Sagar Jatropha Oil Extractions Pvt. Ltd., Emami Group etc. are actively working in biodiesel production in India [8]. Currently, different central level agricultural development organizations and banks provide financial support for the cultivation, processing, production, storage, and distribution of biodiesel producing oilseeds. Presently around 20 biodiesel plants in India are annually producing between 140 to 300 million liters, which is mostly being used to meet the local demand such as irrigation and electricity generation or by automobile and transportation companies for experimental projects. In recent research, it has been found that microalgae can produce higher oil yield with less amount of feedstock. The algae can easily be cultivated on a large scale over a short time period. These algae can grow in different climatic conditions, even when there is very less amount of nutrients available [9].

2.2 Potential Production and Consumption

Indian biofuel policy established in 2018 encourages the use of wastelands for feedstock generation. India has 5,57,665.51 sq. km of wastelands during the year 2015-16 out of 32,87,263 sq. km of the total geographical area [10]. Most of the wastelands can be actively used in non-edible oilseed cultivation exclusively for biodieselproduction. The biofuel policy 2018 also emphasizes on the participation from local communities from Gram Panchayats (local assembly) and Talukas (an administrative district) in planting non-edible oilseed-bearing trees and crops such as *Pongamia Pinnata* (Karanja), *Melia Azadirachta* (Neem), castor, *Jatropha Carcus, CallophylumInnophylum, SimaroubaGlauca,* and *Hibiscus Cannabbinus* for augmenting indigenous feedstock supply for biodiesel production.

India has already set up a goal for creating an additional carbon sink of 2.5 - 3 billion tonnes of CO_2 equivalent by increasing the forest and tree cover by 2030. Energy plantation in the wastelands for production of biodiesel will give a boost to this goal and as well as improve the biodiesel production capacity of the country. India has also planned to decrease the emissions intensity of its GDP by 33 to 35 % by 2030 from 2005 levels. This also can be implemented by reducing dependency upon the positive carbonpetroleum to carbon-neutral biodiesel and thus providing a sustainable energy source.

3. BIOGAS

Biogas is produced by anaerobically digesting organic materials or biomass in a specially designed plant. The biomass can be agricultural waste, kitchen waste, poultry droppings, cattle waste or even municipal waste. As the name suggests, the anaerobic digestion process is achieved in an environment lacking oxygen. The digestion process begins with the hydrolysis process of the feedstocks with the help of bacteria, such as methane bacterium. In this process, the large insoluble biopolymers such as carbohydrates and proteins are broken down into smaller molecules such as sugars and amino acids. Another type of bacteria known as acidogenic bacteria then converts these sugar and amino acid into carbon dioxide, ammonia, hydrogen, organic acids etc. Though a next process known as acetogenesis, these newly developed organic acids are further converted into acetic acid, ammonia, hydrogen and carbon dioxide. Finally, another group of bacteria known as methanogens convert these products into carbon dioxide, methane, hydrogen sulfide etc. The process of biogas production is shown in Fig. 1.1.

The biogas digester for the production of biogas may be designed as a continuous or batch type based upon the availability and volume of feedstock. The calorific value of biogas is around 5000 kcal/m^3. The further advantage of biogas is that the digested slurry from the biogas plant can be used as manure for organic farming.

3.1 Present Status and Future Scope

As shown in Fig. 1.2, India has the second largest production capacity of biogas after China [11]. India has already installed around 50,28,340 biogas plants by the financial year 2018-19 [1].

Recent Trends and Future Prospects of Bioenergy Production in India

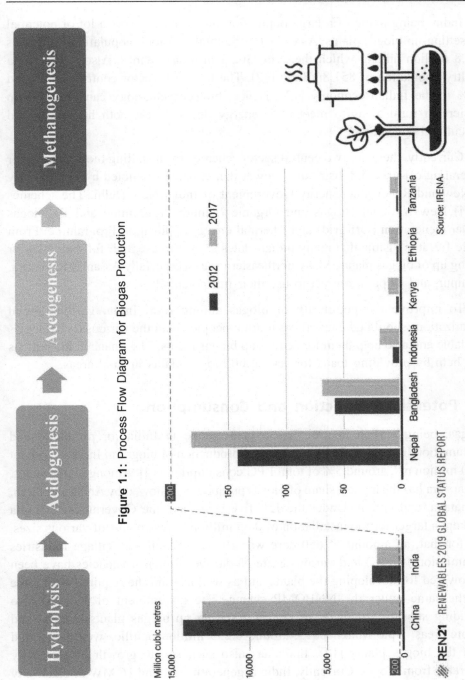

Figure 1.1: Process Flow Diagram for Biogas Production

Figure 1.2: Countrywise Status of Production of Biogas [11]

India, being home of a large population of livestock, bears a lot of potential for setting-up biogas plants. As of 2019, the total livestock population in India is 535.8 million among which there are 302.3 million bovines. Also, India has a poultry population of 851.8million [12]. The livestock sector contributes around 4.9% to the Indian economy [13]. Hence, biogas technology can greatly help farmers in rural areas to meet their energy demands for both household and agricultural operations.

Currently, there are two central sector schemes for installing biogas plants for generating off-grid and distributed power that is being promoted by the Ministry of New and Renewable Energy Government of India, New Delhi. The schemes are i) New National Biogas and Organic Manure Programme and ii) Biogas Power Generation (Off-grid) and Thermal energy application Programme. From Table 1.2, it is visible that many Indian states are yet to use their full potential for setting up of biogas plants. Many northeastern states, especially Assam, Meghalaya, Manipur, and Tripura, are yet to use their full potential.

To improve the popularity of biogas among rural India, various social organizations should take up efforts to train people about the various technologies available and also help them for setting-up biogas plants. The financial institutions also help by providing loans for setting up biogas plants in rural areas.

3.2 Potential Production and Consumption

Biogas, being carbon neutral, has a great scope in India for productionand consumption.In the year 2014-15, the total production of biogas in India was about 2070 million m^3, around 5% of total LPG consumption [14]. Among all the states, Maharastra has the highest share of biogas production, followed by Andhra Pradesh, Karnataka, and Gujrat. Under the 12th five year plan, the Government of India has kept a target for the installation of 0.65 million biogas plants of various sizes, and for that, an amount of 650 core was allotted. Khadi and Village Industries Commission (KVIC) and various State Nodal Departments/Agencies have been empowered for developing the plants and as well as train the required workforce for the same under the NNBOMP scheme.The government of India is also providing subsidies at different rates for setting up biogas plants in rural and remote areas. Upto March 2015, around 0.253 million families were benefitted from the biogas plants [14]. India has also seen steady growth in electricity generated from biogas. Currently, India is generating around 16 MW of electricity from biogas, which is mostly being generated by off-grid systems for local or cluster consumption [15].

Table 1.2: Estimated Potential Vs. The Installed Number of Biogas Plants in India Up to The Financial Year 2018-19 [1]

State/ Union Territories	Nos. of Biogas Plants		Potential used, %
	Estimated Potential	Installed Number	
Andhra Pradesh	10,65,000	5,55,294	52.14
Arunachal Pradesh	7,500	3,591	47.88
Assam	3,07,000	1,38,423	45.09
Bihar	7,33,000	1,29,905	17.72
Chhattisgarh	4,00,000	58,908	14.73
Goa	8,000	4,226	52.83
Gujarat	5,54,000	4,34,995	78.52
Haryana	3,00,000	62,825	20.94
Himachal Pradesh	1,25,000	47,680	38.14
Jammu & Kashmir	1,28,000	3,195	2.50
Jharkhand	1,00,000	7,823	7.82
Karnataka	6,80,000	5,03,935	74.11
Kerala	1,50,000	1,52,019	101.35
Madhya Pradesh	14,91,000	3,73,037	25.02
Maharashtra	8,97,000	9,18,201	102.36
Manipur	38,000	2,128	5.60
Meghalaya	24,000	10,659	44.41
Mizoram	5,000	5,838	116.76
Nagaland	6,700	7,953	118.70
Odisha	6,05,000	2,71,656	44.90
Punjab	4,11,000	1,83,835	44.73
Rajasthan	9,15,000	72,132	7.88
Sikkim	7,300	9,044	123.89
Tamil Nadu	6,15,000	2,23,618	36.36
Telangana	-	19,694	-
Tripura	28,000	3,688	13.17
Uttar Pradesh	19,38,000	4,40,385	22.72
Uttarakhand	83,000	3,63,615	438.09
West Bengal	6,95,000	972	0.14
A&N Islands	2,200	97	4.41
Chandigarh	1,400	169	12.07
Dadra & Nagar Haveli	2,000	681	34.05
Delhi/ New Delhi	12,900	578	4.48
Puducherry	4,300	17,541	407.93
Total	1,23,39,300	50,28,340	40.75

4. BIO-OIL AND GASIFIER

Gasification of biomass is a process for converting solid biomass into gaseous fuel form through a thermochemical process. Whereas, in bio-oil production, the output is in liquid form. The gaseous fuel produced in the gasification process is known as Producer gas. The average composition of producer gas is Carbon Dioxide, Oxygen, Carbon Monoxide, Hydrogen, Methane, and Nitrogen. Producer gas is also sometimes known as gas water gas. The main constituent of the gas, which improves the calorific value of the producer gas is Hydrogen, Methane, and Carbon Monoxide. The concentration of hydrogen in the gas in the producer gas generated from the gasification process mainly depends upon the moisture content, type, and composition of biomass, operating conditions, and configuration of the biomass gasifier [16]. The hydrogen from the producer gas can be separated to be used in fuel cells, or the gas can be used as such along with the producer gas for heat energy production. The calorific value of producer gas is around 1000 – 2000 kcal/m^3.

Gasifier is a device used for gasification and bio-oil production from solid biomass. Inside the gasifier, the feedstock is rapidly heated in the absence of oxygen. The various processes involved in gasification inside the gasifier are drying, pyrolysis, oxidation, and reduction. In the pyrolysis process, the carbonous material is converted into char, tar and oils, and gas. The oils and gases have to be filtered from the tar and char to be used in an engine or other heating applications.

4.1 Present Status and Future Scope

Power from biomass gasification is one of the most important sources of non-conventional power sources in India. As of the financial year 2015-16, India has installed capacity of 4,946.41 MW grid-connected gasification units and 994.46 MW off-grid gasification units. Among the total grid-connected power generation capacity, more than 4,830 MW is generated from bagasse cogeneration, remaining around 115 MW is generated from waste to energy power plants. The off-grid capacity of India comprisesof 652 MW non-bagasse cogeneration, the majority of which are working as captive power plants, in rural areas for meeting the local electrical energy demand,around 18 MW biomass gasifier systems are being used. Around 164 MW biomass asification systems are used in India for thermal applications in industries [17]. The State-wise current capacity of biomass power and cogeneration is shown in Fig. 1.3. In many northwestern states of India, biomass from agricultural crop residues is being burned in the open area and creates smog and air pollution in entire northern India every year. The biomass gasification industry may tap this potential for using this waste residue to generate power and in turn, increase the energy security of those areas.

4.2 Potential Production and Consumption

India, being an agriculture-based economy, holds a huge potential for the production of energy from biomass gasification. Currently, India has a Biomass power potential of 17,536 MW, 5,000 MW from bagasse-based cogeneration in sugar mills, and 2,554 MW from waste to energy units [18]. The current potential for renewable power is shown in Table 1.3.

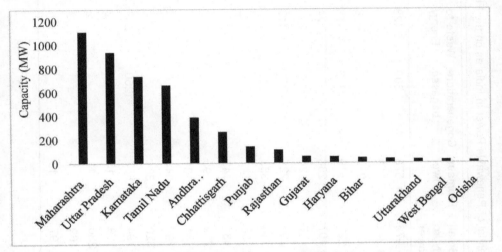

Figure 1.3: State-Wise Biomass Power and Cogeneration Projects [17]

As the demand for energy in India is increasing rapidly due to rapid industrialization, there is a need for tapping into the unused potential of the biomass gasification technologies and implement them on a larger scale.

5. SUSTAINABLE DEVELOPMENT

India, being one of the largest energy consumers of the world, will need a sustainable source of energy in the near future. Petroleum products are the major sources of pollution all over the world, and their reserve is also decreasing at a high rate throughout the world. Also, India has to import quite a large chunk of her domestic consumption from middle eastern countries. The instability in the prices and those countries are affecting the economy of India to a considerable extent and also draining India's forex reserve at a huge rate. Therefore, the

Table 1.3: Source and Statewise Estimated Potential of Renewable Power in India as on the Financial Year 2017-18 [18]

Sl. No.	States/ UTs	Wind Power @ 100m	Small Hydro Power	Biomass Power	Cogeneration-bagasse	Waste to Energy	Solar Energy	Total Estimated Reserves	Distribution (%)
1	Andhra Pradesh	44,229	978	578	300	123	38,440	84,648	7.72
2	Arunachal Pradesh		1,341	8			8,650	10,000	0.91
3	Assam		239	212		8	13,760	14,218	1.30
4	Bihar		223	619	300	73	11,200	12,415	1.13
5	Chhattisgarh	77	1,107	236		24	18,270	19,714	1.80
6	Goa	1	7	26			880	913	0.08
7	Gujarat	84,431	202	1,221	350	112	35,770	1,22,086	11.14
8	Haryana		110	1,333	350	24	4,560	6,377	0.58
9	Himachal Pradesh		2,398	142		2	33,840	36,382	3.32
10	Jammu & Kashmir		1,431	43			1,11,050	1,12,523	10.27
11	Jharkhand		209	90		10	18,180	18,489	1.69
12	Karnataka	55,857	4,141	1,131	450		24,700	86,279	7.87
13	Kerala	1,700	704	1,044		36	6110	9,595	0.88
14	Madhya Pradesh	10,484	820	1,364		78	61,660	74,406	6.79
15	Maharashtra	45,394	794	1,887	1250	287	64,320	1,13,933	10.39
16	Manipur		109	13		2	10,630	10,755	0.98
17	Meghalaya		230	11		2	5,860	6,103	0.56
18	Mizoram		169	1		2	9,090	9,261	0.84
19	Nagaland		197	10			7,290	7,497	0.68

[Table Contd.

Contd. Table]

Sl. No.	States/ UTs	Wind Power @ 100m	Small Hydro Power	Biomass Power	Cogeneration-bagasse	Waste to Energy	Solar Energy	Total Estimated Reserves	Distribution (%)
20	Odisha	3,093	295	246		22	25,780	29,437	2.69
21	Punjab		441	3,172	300	45	2,810	6,768	0.62
22	Rajasthan	18,770	57	1,039		62	1,42,310	1,62,238	14.80
23	Sikkim		267	2			4,940	5,209	0.48
24	Tamil Nadu	33,800	660	1,070	450	151	17,670	53,800	4.91
25	Telangana	4,244					20,410	24,654	2.25
26	Tripura		47	3		2	2,080	2,131	0.19
27	Uttar Pradesh		461	1,617	1,250	176	22,830	26,333	2.40
28	Uttarakhand		1,708	24		5	16,800	18,537	1.69
29	West Bengal	2	396	396		148	6,260	7,202	0.66
30	Andaman & Nicobar	8	8				0	16	0.00
31	Chandigarh					6	0	6	0.00
32	Dadar & Nagar Haveli						0	0	0.00
33	Daman & Diu						0	0	0.00
34	Delhi					131	2,050	2,181	0.20
35	Lakshadweep	8					0	8	0.00
36	Puducherry	153				3	0	156	0.01
37	Others (Industrial waste)					1,022	790	1,812	0.17
	All India Total	3,02,251	19,749	17,536	5,000	2,554	7,48,990	10,96,081	100.00
	Distribution (%)	27.58	1.80	1.60	0.46	0.23	68.33	100.00	

Indian government should actively look for a suitable sustainable energy source for securing the future energy demands of the country.

Bio-energy or energy from biomass is a good option for sustainable energy sources as India already has a huge landmass covered by forest, and also, the agriculture sector is producing a lot of waste materials. Indian government should take the initiative to use those domestic energy sources to get a sustainable and carbon-neutral energy future.

6. CONCLUSIONS

In the current times, a lot of attention has been devoted to the conversion of biomass into fuel for heat generation, transport, electricity generation etc. Biodiesel is considered one of the cleanest liquid fuel alternative to fossil fuels. Bioenergy is an important source of energy for developing countries, mainlyfor the traditional and rural sectors of the economy. The various form of bioenergy production such asbiodiesel, biogas, bio-oil, and gasifier. All these sources of bioenergy are known as a carbon-neutral energy source to fight the build-up of atmospheric carbon dioxide responsible for global warming.The potential bioenergyin India is tremendously high and driven largely by overpopulation and vast agricultural pastures. India is one of the biggest economies with a growing population, big capacities of field and plantation biomass, industrial biomass, forest biomass, urban waste biomass etc. Bioenergy so far is especially prominent in rural India since agricultural residues such as straw and cow dung are easily available. India's import dependency on crude is remarkable i.e. 82% to fulfill the domestic consumption demand, which makes this susceptible to price shocks due to unforeseen escalations in crude oil prices. Hence, bioenergy production would provide energy security for large segments of society, reducing greenhouse gas emissions, and also reducing air pollutants in existing power plants. Indeed, the possibility of a novel bioenergyindustry and the refinery is becoming a thing of reality.

REFERENCES

1. Bio Energy, MNRE. from Ministry of New and Renewable Energy, GoI: https:// mnre.gov.in/bio-energy/current-status. [Accessed on July 23 2020].
2. Biomass Energy in India. from Denmark Embassy in India: https://indien.um.dk /en /innovation/sector-updates/renewable-energy/biomass-energy-in-india/. [Accessed on July 18 2020].

3. Consumption of Petroleum Products. from Petroleum Planning and Analysis Cell, Ministry of Petroleum and Natural Gas, Government of India: https://www.ppac . gov. in/content/147_1_ConsumptionPetroleum.aspx. Accessed on July 20 2020

4. Sheehan J, Camobreco V, Duffield J, Graboski M, Shapouri H. An Overview of Biodiesel and petroleum Diesel Life Cycles. Colorado: National Renewable Energy Laboratory. May 1998.

5. Import/Export. from Petroleum Planning & Analysis Cell, Ministry of Petroleum and Natural Gas, Government of India: https://www.ppac.gov.in/content /212 _1 _ Import Export.aspx.[Accessed on July 25 2020].

6. Motor Vehicles - Statistical Year Book India 2017. from Ministry of Ststistics and Programme Implementation, Government of India: http://mospi.nic.in/statistical-year-book-india/2017/189. [Accessed on July 28 2020].

7. Kumar S, Chaube A, Jain SK. Critical review of jatropha biodiesel promotion policies in India. Energy Policy. 2012; 775-781. https://doi.org/10.1016 /j.enpol. 2011. 11.044.

8. Dewangan A, Yadav AK, Mallick A. Current scenario of biodiesel development in India: prospects and challenges. Energy Sources. 2018; 2494-2501.

9. Khan S, Siddique R, Sajjad W, Nabi G, Hayat KM, Duan P, Yao L. Biodiesel Production From Algae to Overcome the Energy Crisis. HAYATI Journal of Biosciences. 2017; 163-167. https://doi.org/10.1016/j.hjb.2017.10.003.

10. Rao P. Wastelands Atlas of India 2019. Hyderabad: Department of Land Records, Government of India.

11. Renewables 2020 Global Status Report. Paris: REN21.

12. Livestock population in India by Species from National Dairy Development Board: https://www.nddb.coop/information/stats/pop. [Accessed on July 20 2020].

13. National Dairy Development Board from Share of Agriculture & Livestock Sector in GDP: https://www.nddb.coop/information/stats/GDPcontrib. [Accessed on July 17 2020].

14. Abhishek S. Biogas production in India is equivalent to 5% of the total LPG consumption. Retrieved from factly.in: https://factly.in/biogas-production-in-india-is-about-5-percent-of-the-total-lpg-consumption/. [Accessed on July 20 2020].

15. Biogas energy capacity in India from 2009 to 2019 from statista.com: https://www.statista.com/statistics/1044652/india-biogas-energy-capacity/. [Accessed on July 30 2020].

16. Sheth PN, Babu BV. Production of hydrogen energy through biomass (waste wood) gasification. International Journal of Hydrogen Energy. 2010; 10803-10810. https://doi.org/10.1016/j.ijhydene. 2010.03.009.
17. Overview of biomass power sector in India from Biomass Portal, Clean, Green and Sustainable Energy: https://biomasspower.gov.in/About-us-3-Biomass%20 Energy% 20scenario-4.php. [Accessed on July 21 2020].
18. Energy Statistics New Delhi: Central Statistics Office, 2019.

CHAPTER - 2

BIOGAS PRODUCTION, UTILIZATION AND ENTREPRENEURSHIP OPPORTUNITIES

Pradip D. Narale*, Sudhir N. Kharpude and Mahendra S. Seveda

*Department of Renewable Energy Engineering
College of Agricultural Engineering and Post Harvest Technology
Central Agricultural University, Ranipool, Gangtok-737135, Sikkim, India
Corresponding Author

Anaerobic digestion is a simple and low cost process which can be economically carried out in rural areas where organic waste is available in plenty. Animal dung, agriculture waste, municipal solid waste and industrial organic wastes etc. are used as a feedstock for biogas production. Biogas technology is carbon neutral and has positive environmental impact. Biogas not only provides useful energy for cooking, lighting, electricity generation and vehicular application but it also provides organic fertilizer for agriculture hence contributes to sustainable agriculture development and environment protection. The technology solves waste management problem, create hygienic surrounding environment and contributes to socio economic development. In this chapter, biogas production, classification of family size biogas plants, designs of approved models of family size biogas plants in India, entrepreneurship opportunities in implementing small size biogas plants in India, Environmental aspect and its present status will be covered.

1. INTRODUCTION

Biogas can be produced by anaerobic digestion of organic waste material. Anaerobic digestion is a simple and low cost process which can be economically carried out in rural areas where organic waste is available in plenty. Anaerobic digestion is a series of biological processes in which microorganisms break down biodegradable material in the absence of oxygen and produces biogas. The digestion process begins with bacterial hydrolysis of the input materials in order to break down insoluble organic polymers such as carbohydrates, fats, sugar and make them available for other bacteria. Acidogenic bacteria then convert the sugars and amino acids into carbon dioxide, hydrogen, ammonia, and organic acids. Acetogenic bacteria then convert these resulting organic acids into acetic acid, along with additional ammonia, hydrogen, and carbon dioxide. Finally, methanogens convert these products to methane and carbon dioxide. The produced biogas consists of composition of methane, carbon dioxide, hydrogen sulphide, and traces of other gases. Methane is main combustible gas in biogas and can be used as energy source. Figure 2.1 shows the stages of biogas production process.

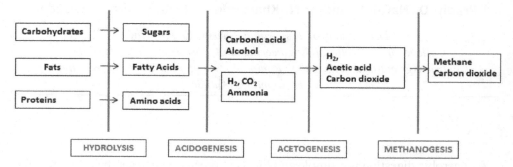

Figure 2.1: Stages of Biogas Production Process

Biogas is also known as the swamp gas, sewer gas, fuel gas, marsh gas, wet gas, and in India more commonly as 'gobar gas'. It consists of 55-60% of methane, 30-40 % of carbon dioxide and trace amount of other gases such as hydrogen sulphide [1,2,3]. The technology of biogas production is well developed and commercially utilized in every parts of world. Small scale family size biogas plants of capacity $1\ m^3$ to $6\ m^3$ are used for household applications such as cooking and lighting purpose whereas large scale biogas plants are used for electricity generation and Bio-CNG production. Animal dung, agriculture waste, municipal solid waste and industrial organic wastes etc. are used as a feedstock for biogas production. Biogas technology is carbon neutral and has positive environmental impact. It avoids addition of extra CO_2 in the atmosphere which helps to control global warming effect. Upgraded biogas can be utilized as an alternate source for vehicular fuel with replacement of conventional vehicular fuels.

Vehicular emission of greenhouse gases can be completely stopped by promoting green Bio-CNG (upgraded biogas) as a vehicular fuel. Biogas not only provides useful energy for cooking, lighting, electricity generation and vehicular application but it also provides organic fertilizer for agriculture hence contributes to sustainable agriculture development and environment protection. The technology solves waste management problem, create hygienic surrounding environment and contributes to socio economic development.

2. FACTORS AFFECTING BIOGAS PRODUCTION

2.1 Waste Composition

The substrate composition is major and determining factor in anaerobic digestion process. The wastes treated by anaerobic digestion may comprise a biodegradable organic fraction and lignocellulosic organic materials. The biodegradable organic fraction includes kitchen scraps, food residue, and grass and tree cuttings and lignocellulosic organic matter containing coarser wood, paper, and cardboard besides plastics. The more the biodegradable organic fraction in waste higher will be the gas yield. Most literature sources report substantial decrease in degree of polymerization of lignocellulosic organic fraction by weaking the bond between lignin and carbohydrate using several pretreatments as mechanical size reduction, heat treatment, chemical treatment and biological treatments.

2.2 Volatile Solid

Volatile solids are the portion of solids in biodegradable organic fraction that has calorific value. In anaerobic digestion volatile solids are broadly associated with digestable fraction of biodegradable organic waste, and depending upon input nearly one-third to one-half of biodegradable organic fraction are volatile solids which are converted into biogas. In fact, biogas yield are often reported by researchers in cubic meter per mass of volatile solids. Thus wastes characterized by high volatile solids are best suited for anaerobic digestion system.

2.3 Alkalinity and pH

The methane production efficiency is the most important evaluation yardstick in anaerobic digestion, there are several factors affecting methane production efficiency such as pH, temperature, type and quality of substrate, mixing etc., and the value of pH is the pivotal factor [4]. Sufficient alkalinity is essential for pH control. Alkalinity serves as a buffer that prevents rapid change in pH. Alkalinity results from release of amino groups and production of ammonia as proteinaeceous wastes are degraded.

Ammonium ion acts as buffer in the anaerobic reactor. Anaerobic bacteria, specially the methanogens, are sensitive to acid concentration within the digester and their growth can be inhibited by acidic conditions. It has been determined that an optimum pH value for anaerobic digestion lies between 5.5 and 8.5. But once methane production is stabilized, the pH level stays between 7.2 and 8.2 [5].

2.4 Volatile Fatty Acids Concentration

Volatile fatty acids, usually made up of chains of carbon molecule are formed in a process of fermentation [6]. Volatile fatty acids are important intermediate compounds in the metabolic pathway of methane fermentation and cause microbial stress if present in high concentrations. The intermediates produced during the anaerobic bio-degradation of an organic compound are mainly acetic acid, propionic acid, butyric acid, and valeric acid. Amongst these, acetic and propionic acids are the major volatile fatty acids present during anaerobic bio-degradation and their concentrations provide a useful measure of digester performance [5].

2.5 Temperature

Due to the strong dependence of temperature on digestion rate, temperature is the most critical parameter to be maintained in the desired range. There are two temperature ranges that provide optimum digestion conditions for the production of methane i.e. the mesophilic and thermophilic ranges. The optimum temperature for mesophilic digestion is 35°C and a digester must be maintained between 30°C and 35°C for most favorable functioning. The thermophilic temperature range is between 50°C-65°C. It has been observed that higher temperatures in the thermophilic range reduce the required retention time.

Thermophilic digestion allows higher loading rates and achieves a higher rate of pathogen destruction besides higher degradation of the substrate. The reaction is however, sensitive to toxins and smaller changes in the environment and requires better control. The microbial growth, digestion capacity and biogas production are enhanced by thermophilic digestion, since the specific growth rate of thermophilic bacteria is higher than that of mesophilic bacteria [5].

2.6 Carbon to Nitrogen Ratio (C/N ratio)

The relationship between the amount of carbon and nitrogen present in organic materials is represented by the C/N ratio. Carbon is energy producing factor while nitrogen is required for producing new cell mass [7]. A nutrient ratio of the elements C: N: P: S at 600:15:5:3 is sufficient for mechanization [8].

Optimum C/N ratios in anaerobic digesters should be between 20–30 in order to ensure sufficient nitrogen supply for cell production and the degradation of the carbon present in the wastes. As the reduced nitrogen compounds are not eliminated in the process, the C/N ratio in the feed material plays a crucial role [7]. A high C/N ratio leads to rapid consumption of nitrogen by methanogens and results in lesser biogas production. On the other hand, a lower C/N ratio causes ammonia accumulation and a pH value exceeds 8.5, which is toxic to methanogenic bacteria. Optimum C/N ratios of the digester materials can be achieved by mixing materials of high and low C/N ratios, such as organic solid waste mixed with sewage or animal manure [5].

2.7 Hydraulic Retention Time (HRT)

The hydraulic retention time is the ratio of the digester volume to the influent substrate flow rate. It is a measure of the time the substrate resides inside the digester. The required retention time for completion of the anaerobic digestion reactions varies with differing technologies, process temperature, and waste composition. Retention time is an important operational variable which can be easily manipulated according to needs. It is possible to feed sufficient waste in anaerobic digester by decreasing retention time [5,9].

2.8 Organic Loading Rate (OLR)

Low solids anaerobic digestion systems contain less than 15 % total suspended solids. High solid processes have about 20% or higher total suspended solids. An increase in total solids in the reactor results in a corresponding decrease in reactor volume. The organic loading rate is a measure of the biological conversion capacity of the anaerobic digestion system. The feeding system above its sustainable organic loading rate results in low biogas yield due to accumulation of inhibiting substances such as fatty acids in the digester slurry. In such a case, the feeding rate to the system must be reduced. Organic loading rate is a particularly important control parameter in continuous systems. Many plants have reported system failures due to overloading [5].

3. STATUS OF FAMILY SIZE BIOGAS PLANTS IN INDIA

As per Ministry of New and Renewable Energy (MNRE), Government of India, New Delhi statistics, India has installed around 50, 28, 347 (Approx. 50. 28 Lakh) small scale family size biogas plants by the end of 31st March 2019 through National Biogas Manure Management Programme (NBMMP) from year 1981-82 to 2017-18 and New National Biogas and Organic Manure Programme (NNBOMP) from year 2018-19 to 2019-20. In 2019-2020 country has set up target to install around 70,000 biogas plants in different states and union territories of India through New National Biogas and Organic Manure Programme

(NNBOMP). Out of set up target country has installed only 12,019 family size biogas plants through NNBOMP by the end of December 2019.

4. CLASSIFICATION OF BIOGAS PLANT

Based on nature of application, biogas digesters can be broadly classified into three sub groups: family size (1-10 m^3/d), medium scale (1000-5000 m^3/d) and large scale (>5000 m^3/d). Based upon the nature of flow maintained in the biogas digester, digester can be classified into two groups: The batch type and Continuous type digester.

a. Bach Type digester has simplest design and easy to operate. The feed material is loaded once and changes are allowed to take place and biogas is produced. However, the biogas production varies considerably with time. In such type of digester operation, gas production follows sigmoid curve where after reaching at peak, gas production starts decreasing continuously.

b. Continuous flow type digesters have nearly constant feed rate, slurry discharge and gas production. At start-up these digester behave as batch digesters but after considerable time steady state is achieved. Like batch digesters they too follow sigmoid curve, but once gas production reaches at peak, it became constant until and unless there is any problem with the process or feed rate is lower down. Most of the industrial scale digesters are continuous flow type.

4.1 FAMILY SIZE BIOGAS PLANTS

4.1.1 Fixed Dome Type Biogas Plant (Deenbandhu Biogas Plant)

Deenbandhu is improved version of Janta fixed dome biogas plant developed by an NGO named Action for Food Production (AFPRO) in 1984. This model is 30 per cent cheaper than Janta model and 45 per cent cheaper than KVIC plants of comparable capacities. Deenbandhu model is 15-30 percent cheaper than Pragati model developed by BORDA and UNDARP (United Socio Economic Development of Research Programme). Deenbandhu model has curved bottom and a hemispherical top which are joined at their bases with no cylindrical portion in between. An inlet pipe connects mixing tank with the digester. Cow dung slurry in 1:1 ration with water is prepared in mixing tank and fed up to the level of second step in the outlet tank which is also the base of the outlet displacement chamber (Fig. 2.2). Generated biogas accumulates in the empty portion of the plant which creates pressure on slurry and displaces it into the outlet displacement chamber. Design of Deenbandhu biogas model of capacity 1, 2, 3, 4 and 6 m^3 was approved by MNRE, Govt. of India in 1986 for countrywide popularization under the NPBD.

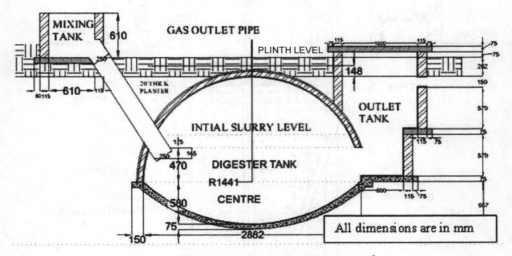

Figure 2.2: Deenbandhu Biogas Plant ($2m^3$)

4.1.2 Floating Dome Type Biogas Plant (KVIC Biogas Plant)

KVIC biogas plant was developed by Khadi and Village Industrial Commission (Fig. 2.3 and Fig. 2.4). It basically comprises of cylindrical digester made up of cement concrete structure with attached inlet and outlet connections and floating dome for gas collection made of mild steel material. Gasholder moves up and down guided by a central guide pipe depending upon accumulation and discharge of gas. The initial cost of installation of this biogas plant is higher than fixed dome type plant and also involved higher cost of repair and maintenance. Movable gasholder made of mild steel alone accounts for 40 per cent of the total plant.

4.1.3 Prefabricated Biogas Plants

The digester of the plant is made of Reinforced Cement Concrete (RCC) whereas KVIC floating dome for gas storage is made of FRP/HDPE material. The cost of this biogas plant is lower than the existing KVIC biogas plant. It is necessary to keep some weight over the surface of prefabricated FRP/HDPE floating dome of the plant to create constant pressure of biogas at the inlet of burner. Figure 2.5 shows pictorial view of the prefabricated biogas plant with bricks over the surface of floating dome to get required biogas pressure.

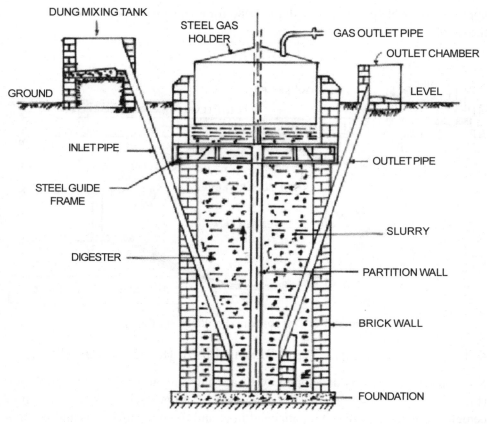

Figure 2.3. Schematic View of KVIC Biogas Plant

Figure 2.4. Pictorial View of KVIC Biogas Plant

Figure 2.5: Pictorial View of Prefabricated Biogas Plant

4.1.4 Bag Type Biogas Plants (Flexi model)

Bag type biogas models are simple in construction, easily understandable mechanics and made up of polyethylene material. Flexi model consist of horizontal bag type structure with inlet and outlet PVC pipe attachments. Due to flexible bag type of structure biogas plant has lower life as compared to fixed dome and floating drum biogas plant. The schematic and pictorial view of the flexible bag type biogas plant is given in Fig. 2.6 and Fig. 2.7 respectively.

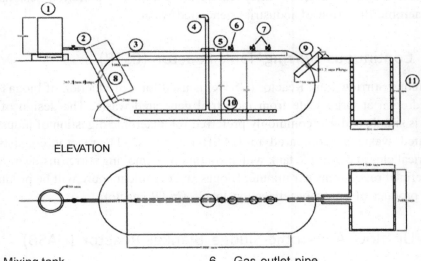

1	Mixing tank	6	Gas outlet pipe
2	PVC ball pipe	7	Secondary gas outlet pipes
3	Flexible balloon	8	Inlet pipe
4	Stirring unit inlet	9	Outlet pipe
5	Safety valve	10	PVC piping system with fine perforations
		11	Slurry pit

Figure 2.6: Schematic Diagram of Flexible Ballon Biogas Plant

Figure 2.7: Picvtorial View of Flexible Ballon Biogas Plant

4.2 Industrial Biogas Plants

Waste residues generated from the industries in the form of solids or liquid form will be useful as a feedstock for an anaerobic digestion in specifically designed industrial biogas plant. Continuous Stirring Tank Reactor (CSTR) and Up flow Anaerobic Sludge Balnket (UASB) reactor are commonly preferred designs for the anaerobic digestion of industrial generated wastes.

4.2.1 Continuous Stirring Tank Reactor (CSTR)

Continuous Stirring Tank Reactor (CSTR) is used for a production of biogas and bio hydrogen at large scale from industrial generated waste. The design of the CSTR is simple and it is commonly preferred for anaerobic digestion of industrial generated waste as compared to UASB reactor. CSTR reactor consists of cylindrical closed anaerobic tank with a continuously moving stirrer inside moving with defined revolutions per minute. Biogas and organic manure will be produced as the product of anaerobic digestion inside CSTR reactor.

4.2.2 Up flow Anaerobic Sludge Blanket Reactor (UASB)

Up flow anaerobic sludge blanket reactor is a form of anaerobic digester used for industrial waste water treatment. It uses anaerobic process in a closed tank where a blanket of granular sludge is suspended. Industrial wastewater flows upward through this sludge blanket and is processed by the anaerobic bacteria. Biogas along with organic solid liquid fertilizer in the form of digested slurry will be generated due to the action of anaerobic digestion. UASB reactors typically suitable for a dilute waste streams up to 3% TSS (Total Soluble Solids).

5. APPROVED MODELS OF FAMILY TYPE BIOGAS PLANTS

Ministry of New and Renewable Energy (MNRE), Government of India, New Delhi has approved various family size biogas plant models of 1 m^3 to 6 m^3 per day capacity as listed in Table 2.1. The MNRE approved models are eligible for central financial assistance under 'New National Biogas and Organic Manure Programme' (NNBOMP). Ministry of New and Renewable Energy (MNRE), Government of India, New Delhi can also approve new innovative and cost effective biogas models after its positive testing reports from the Ministry's certified testing centers.

Table 2.1: List of MNRE Approved Models of Family Size Biogas Plants 1 m³ to 6 m³ per Day Capacity [10]

S. No.	Biogas Plant Models
1.	**Fixed Dome Biogas Plants:**
	(i) Deenbandhu fixed dome model with Brick masonry construction.
	(ii) Deenbandhu ferro-cement model with in-situ technique.
	(iii) Prefabricated RCC fixed dome model.
2.	**Floating Dome Design Biogas Plants:**
	(i) KVIC floating steel metal dome with brick masonry digester.
	(ii) KVIC floating type plant with Ferro-Cement digester and FRP gas holder.
	(iii) Pragati Model Biogas Plants.
3.	**Prefabricated model Biogas Plants:**
	(i) Prefabricated Reinforced Cement Concrete (RCC) digester with KVIC floating drum.
4.	**Bag Type Biogas Plants (Flexi model)**

6. HURDLES IN IMPLEMENTING FAMILY SIZE BIOGAS PLANTS IN INDIA

6.1. Availability of Feedstock

Animal waste generally used as feedstock for family size biogas plants. Availability of animal dung throughout year is necessary for functioning biogas plants. Lack of availability of animal dung is the bigger constraint for implementation of family size biogas plant.

6.2 Economic Constraints

Because of small land holding area and variable climate, economic condition of the farmers in India is very poor. Therefore most of the farmers are not able to manage the initial capital investment incurred in construction of this family size biogas plants.

6.3 Promotion of Technology

One of the major hindrances in the adaptation of biogas technology is the lack of awareness among farmers. Many farmers in the rural areas have feedstock resources for biogas production, but due to lack of promotional activities they are not aware with the biogas technology, benefits, maintenance and respective government schemes. Although eight BDTCs in India are working hard to reach mass of people through exhibitions and social campaigns but till now huge mass of rural peoples are untouchable. Therefore it is necessary to gear up promotional activities to create awareness and speed up implementation of family size biogas plants to achieve target.

6.4 Availability of Implementing Agency or Skilled Labor at Root Level

Availability of skilled labor and implementing agency is always big constraint for implementation of biogas plants in rural areas of country.

6.5 Social Dilemma

There is social dilemma to handle and cook food on biogas generated from cow dung and human waste. People do not prefer to cook food on biogas generated from human waste.

7. ENTREPRENEURSHIP OPPORTUNITIES IN IMPLEMENTING FAMILY SIZE BIOGAS PLANTS IN INDIA

Biogas Development and Training Centres (BDTCs) provide biogas plant development and maintenance training to entrepreneurs and masons. These trained entrepreneurs and masons construct family size biogas plant after identification of beneficiary and complete all documental requirements as needful to avail subsidy. The country installed around 52.28 lakh number of family biogas plants till 31st

March 2019 and has total potential to implement around 125 lacks of family size biogas plants. MNRE, Govt. of India has set up target to install 76,000 small family size biogas plants for the year 2019-2020, which was very low if we consider the total potential of India. Entrepreneurs have tremendous scope here to gear up the speed of implementation of family size biogas plant in rural areas of India.

Apart from existing approved biogas models entrepreneurs can develop and implement their own biogas model either it may be prefabricated, bag type, or cemented fixed or floating biogas plant after testing and MNRE approval. Entrepreneurs can also sell their own unapproved biogas models to the needy farmers without claiming subsidy. MNRE Government of India provides subsidy to only approved models of biogas plants after its implementation and commissioning.

8. APPLICATION OF BIOGAS

Biogas being a clean gaseous renewable fuel can be utilized in several ways. Most commonly it is used for cooking and lighting applications in many developing countries. In recent years application of biogas is attempted in several other purposes like power generation, transportation fuel, refrigeration and cogeneration etc.

8.1 Biogas used as a Cooking Fuel

Biogas provides clean and efficient fuel for cooking. Biogas burner comprises of nozzle, an air inlet, a mixing chamber and fire sieve element. Biogas stoves normally operate at gas pressure of 75 to 90 mm water column. Brightness and combustibility of gas can be controlled by regulating gas pressure and air fuel ratio. Fuel to air ratio generally maintained at 1:10. 0.227 m^3 of biogas is required per day to meet out cooking need of one person.

8.2 Use of Biogas as a Lighting Fuel

Biogas provides clean fuel for lighting homes. Biogas lamp comprises of nozzle, an air inlet, mixing chamber and mantle. Mantle is made of ramine fiber and is coated with thorium nitrate solution. Ramine fiber reduces to ash while burning and forms a layer of thorium dioxide which emits dazzling white light at high temperatures. Brightness of biogas lamp mainly depends on factors like gas pressure, relative proportion of gas and air mixture which is about 1:10 and thoroughness in mixing. Biogas lamps are generally designed to produce 100 candle power and consume 0.125 m^3 biogas per hour.

8.3 Biogas for Power Generation

Biogas is used to generate electric power as decentralizes source of energy in several countries. The technology for conversion of biogas to electrical power is simple and well known. Electric generators which can run on biogas are available in all sizes in virtually all countries for biogas to electricity conversion. It was stated that 1 kWh of electricity can be generated from 0.75 m^3 of gas which can light 25 electric bulbs of 40W rating whereas 0.75 m^3 of biogas if directly burnt can light only 7 biogas lamps for one hour [11]. Popularly, there are three technologies to convert the conventional engine to the biogas engine. Table 2.2 shows conversion technologies and their advantages and disadvantages. According to this table, high efficiency biogas engine can be converted from compression ignition engine [12].

Table 2.2: Comparison of Conversion Technology for Biogas Engine [12]

S. No.	Based Engine	Conversion Technology	Advantage	Disadvantage
1.	Spark Ignition Engines	No modification	- Simple and low cost. - Carbon Neutral	- Low engine efficiency
2.	Compression Ignition Engines	Reduce compression ratio, set up ignition system	- Complicated, high cost. - Carbon Neutral	- Medium engine efficiency.
3.		Using dual fuel	- Simple, low cost. - Reduce CO_2.	- Diesel knocking. - Need fossil fuel for operation

8.4 Biogas used as Transportation Fuel

Biogas consists of 55-60% methane, 30-40% of CO_2 and traces of other gases. Methane is main combustible gas in biogas and carbon dioxide is unwanted gas which is incombustible. This unwanted CO_2 can be removed using biogas scrubbing technologies like water scrubbing technology, Pressurize swing adsorption technology, cryogenic up-gradation technology etc. Upgraded biogas which contents 90% methane can be utilized as a source of fuel for transportation after high pressure compression at 200 bar in cylinder. This high pressure methane gas produced from anaerobic digestion of organic waste is commonly known as Bio CNG.

8.5 Biogas used as a Heat Engine

In modern waste management amenities, biogas can be used to run any type of heat engine in order to generate electrical or mechanical power. It can be compressed, like a natural gas, to control motor vehicles.

8.6 Cogeneration

Cogeneration is the combined production of electricity and heat. It is one of the promising technologies to meet energy efficiency targets and to minimize hazardous gas emissions. The purpose of cogeneration is to simultaneous production of electricity along with heat so that maximum quantity of energy can be harnessed per unit of fuel used. The produced electricity can be consumed locally or which can be directly feed to the electricity grid. The produced heat can be used for room heating, domestic hot water and industrial application.

The energy utilization efficiency by cogeneration mode can be increased up to 80-90%. Cogeneration engines are available in various power ranges which vary from few hundreds of kW to 3 MW electric plants. Biomethanation is well adapted to cogeneration in most of the developing countries due to high electricity cost and availability of biological organic waste streams like slaughter houses, sewage plants, food industries, breweries etc. The technology is also well suitable for cold remote areas of India where the grid electricity is always a bottle neck issue.

8.7 Utilization of Biogas for Water Pumping

Biogas is also utilized for water pumping application using biogas run IC engines. 0.445 m^3/hr of biogas is required to run one hp biogas engine.

8.8 Biogas used in Fuel Cell Technology

Biogas can be directly converted in to electricity using fuel cell technology. A fuel cell is an electrochemical device which directly converts chemical energy of conventional or non-conventional fuel into low voltage electricity. However for this direct conversion it requires very clean gas and highly expensive fuel cells. Therefore the technology is still matter of research and not practical option for commercial production of power.

8.9 Biogas used in Refrigeration

Biogas can also be utilized to run vapor absorption refrigeration cycle. Biogas refrigeration system provides economical feasible renewable option for cold storage of horticulture crops to rural India. Technology reduces emission of hazardous greenhouse gases in the atmosphere and helps in climate change mitigation. The refrigeration system helps to reduce post-harvest losses of perishable fruits and vegetables and contribute in food security and nutritional availability.

9. ENVIRONMENTAL IMPACTS AND SUSTAINABILITY

Biogas production is a biochemical system that converts any organic waste stream in to biogas and organic manure. Environmental impacts of a household and large scale biogas system depends upon manure management system and use of the biogas. Biogas production system will have the following benefits,

9.1 Odour

Anaerobic digestion system provides the effective odour reduction. Natural fermentation of wastes results in the production of hydrogen sulphide, ammonia and VFA (Volatile fatty acids). VFA is most responsible substance for odour issues. In anaerobic digester VFA are digested by bacteria which reduce the overall odour emissions from the manure.

9.2 Pathogens

Biogas production system reduces pathogens from the wastes and provides pathogen free organic manure for agriculture application.

9.3 Green House Gases

Biogas production systems reduce harmful greenhouse gas emissions like methane (CH_4) and carbon dioxide (CO_2) in to the environment and contribute in to the sustainable environment.

9.4 Sustainability

The biogas production is sustainable, renewable, carbon neutral and reduces the dependency from fossil fuels like coal, natural gas etc. The carbon neutral biogas system reduces the emission of greenhouse gases in to environment. Beneficiaries of biogas plant can replace the coal, wood, and natural gas by clean alternate biogas fuel to meet their energy need. In addition to the biogas fuel, beneficiaries can get high quality enriched organic fertilizer for their agriculture application. Farmers can replace the use of chemical fertilizers by organic fertilizers in their own agriculture land and can improve the fertility of soil. Farmers can get increased yield and high quality agriculture product by application of organic fertilizers which contribute for the sustainable growth of agriculture. The system can improve human sanitation, waste management, soil fertility and replace the direct use of conventional fossil fuels for the cooking application which directly benefit for the sustainable environment.

REFERENCES

1. Mc Kendry P. Energy production from biomass (part-2): conversion technologies. Bioresource Technology. 2012; 83: 47–54.
2. Rajendran K, Aslanzadeh S, Taherzadeh MJ. Household biogas digesters: a review. Energies. 2012; 5(8): 2911–2942.
3. Hiremath RB, Kumar B, Balachandra P, Ravindranath NH, Raghunandan BN. Decentralised renewable energy: scope, relevance and applications in the Indian context. Energy Sustain Dev. 2009; 13: 4–10.
4. Liu CF, Yuan XZ, Zeng G, Li W. Prediction of methane yield at optimum pH for anaerobic digestion of organic fraction of municipal solid waste. Bioresource Technology. 2008; 99(4): 882-888.
5. Chaudhary KB. Dry continuous anaerobic digestion of municipal solid waste in thermophilic conditions. Thesis- Masters in Engineering, Asian Institute of Technology, Thailand; 2008: 93.
6. Fleming SE, Arce DS. Volatile fatty acids: their production, absorption, utilization, and roles in human health. Clinics in Gastroenterology. 1986; 15(4): 787-814.
7. Miller C. Understanding the carbon-nitrogen ratio. Acres: The voice of Eco-agriculture, 2000; 30(4): 20-21.
8. Puertas J. Triennium Work Report on Renewable gas - The Sustainable Energy Solution, 25th World Gas conference Report, Kuala Lumpur, Malaysia. 2012; 52.
9. Wang Q, Noguchi CK, Kuninobu M, Hara Y, Kakimoto K, Ogawa HI, Kato Y. Influence of hydraulic retention time on anaerobic digestion of pretreated sludge, Biotechnology Techniques. 1997; 11(2):105-108.

10. Ministry of New and Renewable Energy (MNRE), Govt. of India. Annual Report of MNRE, https://mnre.gov.in/img/documents/uploads/file_f-1597797108502.pdf; 2020 [Accessed 22 October 2020].
11. Mittal KM. Biogas Systems: Principles and Applications. 1st ed. New Delhi: New Age International Publishers; 1996.
12. Dung NN. A study of conversion of diesel engine to fully biogas engine with electronically controlled. International Journal of Earth Sciences and Engineering. 2012; 5(6): 1745-1749.

CHAPTER - 3

ADVANCEMENTS IN BIOGAS SLURRY MANAGEMENT TECHNOLOGIES

Madhuri More* and Deepak Sharma

Department of Renewable Energy Engineering
College of Technology and Engineering
Maharana Pratap University of Agriculture and Technology,
Udaipur-313001, Rajasthan, India
*Corresponding Author

Biogas technology is a promising renewable energy technology, which is very useful for the production of biogas and biogas slurry. The biogas slurry is produced from the biogas plant under the process of anaerobic digestion. Biogas slurry having good nutrient contents, which is suitable for crop productivity and reduce the extra use of mineral fertilizers. The overall aim is to secure good quality organic fertilizer and to enhance its use for agricultural purpose. The utilization of biogas slurry has several environmental benefits. Nowadays, the production of biogas slurry is increasing; however, there is need to store. The higher production of biogas slurry would pollute water bodies and environmental issues i.e. nitrogen losses because of improper management of waste. For the proper utilization, biogas slurry processing and up-gradation technologies are available, which helps to reduce transportation, handling and storage cost. A solid-liquid separation technology is used for biogas slurry processing by using modern equipment's. After processing, produced bio-fertilizers used as good quality organic manure, which is easy to marketable. This chapter emphasis the problems related to the utilization and handling biogas slurry as a bio-fertilizer and improving biogas slurry quality. This chapter discuss the problems related to the transportation and handling and advance technologies for the utilization of biogas slurry.

1. INTRODUCTION

Nowadays, energy is an important resource in India. Energy can be divided into two categories, namely, renewable energy and non-renewable energy [1]. Renewable energy is an inexhaustible in nature including solar energy, wind energy, bio-energy, hydropower, ocean and geothermal respectively. Non-renewable energy source is an exhaustible in nature (i.e. Coal, oil and natural gas), which is used to achieve current demand [2]. Few years ago, energy cost has been increased because of due to its increasing energy demand and decreasing conventional fuel availability. In the world, environmental policies are adopted for increasing demand on sustainable renewable energy [3].

Biogas production is source of renewable energy. In Asia, Biogas production technology is developed technology and domestic size of biogas plants are very popular [4]. Biogas technology has used in mitigating change in climate and reducing issue related to global warming due to the wider usage of wood and kerosene. It also contributed in global carbon reduction [5]. There is a tremendous potential in the world as well as in India. In India, biogas plant is considered as a huge potential for biogas production, over 20,125-biogas plants installed in the year 2017-18 [6]. About 2.5 million of biogas plants operate in India by 2020 [3]. These biogas plants leads to the produce biogas and 78 million tonnes of biogas slurry produced per year.

Biogas slurry produced from the biogas plant has considerable amount nitrogen (N), potassium (K) and phosphorus (P); therefore, it is very useful to the crop production as a valuable organic fertilizer [7, 8, 9, 10, 11]. As per the MNRE reported that, biogas slurry produced from biogas plants contains 80% carbon, 1.80 % nitrogen, 1.0 % phosphorus, and 0.90 % potash, it is very healthy source to boost crop production. Biogas slurry is an inexpensive way to reduce the use of chemicals and mineral fertilizers in the soil and hence reduces the environmental pollution [8]. Nowadays, the production of biogas slurry is increasing; however, there is need to store biogas slurry in dry condition. For the proper utilization and handling of biogas slurry, solid-liquid separation is a common method, which helps to reduce transportation, handling and storage problems. There are several techniques for the separation of biogas slurry like screw press, centrifugation, sieving or screening, belt filters, and sedimentation or chemically enhanced sediment settling [10, 12, 13, 14].The separated solid and liquid partitioning divides most of the nitrogen in liquid fraction and phosphorous in solid fraction, which helps to the plant nutrients management for better crop production [14]. The solid fractions having phosphorous content can be use as phosphorous rich manure for improving soil characteristics. The major fraction deriving from the separation is liquid phase, which contains minimum percentage of potash and higher percentage of nitrogen content. The liquid phase of separated biogas slurry used as pesticide and fungicide [15].

2. BIOGAS SLURRY PRODUCTION

For the biogas slurry production, various biogas plants design has been developed. The biogas plants mainly classified into two categories i.e. floating drum type and fixed dome type. The anaerobic digestion process is complete cycle to convert organic matter into biogas slurry (Fig. 3.1).

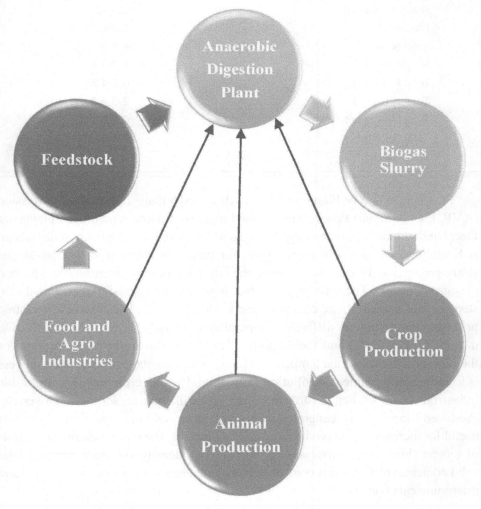

Figure 3.1: Complete Cycle of Biogas Slurry Production

The physiochemical analysis of biogas slurry is presented in Table 3.1.

Table 3.1: Physiochemical Analysis of Biogas Slurry [10, 14]

Parameters	Biogas slurry
pH	8-8.5
Density (kg/m^3)	1200-1350
Electrical conductivity (dS/m)	3.0-3.2
Moisture content (%)	78-92
Total solid content (%)	8-12
Volatile solid content (%)	70-75
Ash content (%)	25-30
DM (%)	5.0-6.5
Total N (kg t^{-1})	4.1-5.5
NH$_4$-N (kg t^{-1})	3.0-3.5
Total P (kg t^{-1})	0.5-1.5
Total K (kg t^{-1})	2.5-4.5
Total Mg (kg t^{-1})	0.50-1.10
Total Cu (kg t^{-1})	9.5-13.0
Total Ca (kg t^{-1})	1.0-3.0

Floating drum type biogas plants are Khadi and Village Industries Commission (KVIC), Ganesh and Pragati biogas plant and fixed dome type biogas plants are Deenbandhu biogas plant and Janta biogas plant. In India, the most popular design is KVIC and Deenbandhu biogas plant for biogas production as well as biogas slurry production. Biogas slurry is produced from the process of anaerobic digestion of organic matters. The anaerobic digestion is process in which different degradation steps such as hydrolysis, acidogenesis and methanogenesis. The anaerobic digestion process carried out at different temperatures such as psychrophilic (<25°C), mesophilic (25-45°C) and thermophilic (45-70°C). The suitable temperature for the digestion is mesophilic temperature [2, 3]. The advantages of digestion process is to reduce environmental pollution from animal waste, crop residues, digestible industrial waste, kitchen waste and energy crops like maize, sorghum, clover etc. Produced biogas slurry generally consists of macro and micronutrients, which is useful for increasing crop productivity and soil quality. The physicochemical analysis of biogas slurry can be assessed in terms of pH, density, moisture content, total solid content, volatile solids content, ash content, electrical conductivity, macro and micronutrients content.

3. APPLICATIONS OF BIOGAS SLURRY

3.1 Organic Manure

The application of biogas slurry is very suitable to improve the physicochemical and biological properties of soil. Biogas slurry contains amounts of nutrients such as macro (NPK) and micronutrients, which help to increase the plant growth [9]. The various researches reported that the use of produced slurry is useful as organic fertilizer, pesticide or fungicide.

3.2 Composting Material

Biogas slurry being 90 to 93 % moisture content and full of bacteria can be utilized as composting material with others organic residues and substrates [9]. Composting material like vegetative material, solid organic fibrous substrates and agricultural wastes are available locally. These materials with biogas slurry inoculating effective earthworms and microorganism could be the suitable for improving quality and enhancing quantity of organic manure. The use of composted biogas slurry is very helpful for long-term improvement of soil fertility.

3.3 Vermicomposting

Biogas slurry has potential for earthworm culturing due to its easily digestible nature and nutrient contents. For vermicomposting process, the final moisture content of substrate is vary from 65 to 75 %. Biogas slurry having higher moisture content so that the some agricultural material adding into the substrate for saving earthworms life. The vermicomposting biogas slurry helps to increase storability and transportability, which is an another advantages on additional income and marketing sector [10].

3.4 Bio-Pesticide Applications

Biogas slurry has been investigated for its pesticide potential for controlling red spider, attacking vegetables, wheat crop and cotton crop. The application of slurry helps to increase quality of crop and as pest control. The application of produced slurry as organic manure is also control the environmental pollution.

3.5 Others Applications

The application of biogas slurry is as carrier for fungal bio-pesticide formulations and helps to increasing biomass production. The use of dehydrated slurry having vitamin B_{12} and nutrients has been used as a feed for poultry bird's fish ponds and pig.

4. UTILIZATION OF BIOGAS SLURRY

Biogas plants produce biogas as well as biogas slurry. Biogas slurry is an excellent crop fertilizer, which is rich in macronutrients and micronutrients. The properties of biogas slurry depend on the composition of raw material and nature for anaerobic digestion of raw material. Biogas slurry is normally used as fertilizer but there is need to store for crop productivity. Traditionally, biogas slurry is utilized in terms of drying, composting, farmyard manure (FYM), vermicomposting. There is main problem related to utilization and handling of biogas slurry. However, there is a need to recycle nutrients from manure for efficient nutrients management [10]. For the environmental point of view, biogas slurry processing and its up-gradation is important for the utilization of slurry waste as organic fertilizer [14].

4.1 Biogas Slurry Processing

Biogas slurry used without any primary treatment. The cost of transport, storage and handling are varying as compared with fertilizer value. The cost increases because of storage capacities. The fertilizer applications period is limited i.e. growing season, so that utilization of biogas slurry is very difficult. For the environmental protection, biogas slurry processing is necessary to transport and distribute nutrients on the field.

Biogas slurry processing carried out on different developed technologies. The solid-liquid separation technology is most acceptable technology for biogas slurry processing. The aim of biogas slurry processing is to produce standardised bio fertilizers. Biogas slurry processing mostly useful for reduce the volume of slurry and solid-liquid separation from biogas slurry (Fig. 3.2).

The partial processing is relatively simple technology and very cheap for processing of biogas slurry. For complete processing, there are several methods are available in which solid-liquid separation methods are suitable for utilization of biogas slurry. Solid-liquid separation techniques such as screw press technology, decanter centrifuge, belt filter and use of precipitation agent are used for separation [10, 12, 13, 14].

4.2 Advanced Technologies for Utilization of Biogas Slurry

4.2.1 Screw Press Separation Technology

Screw press separation technologies are often used in large-scale biogas plant. The separation is carried out in terms of fibre content in the digester. Figure 3.3

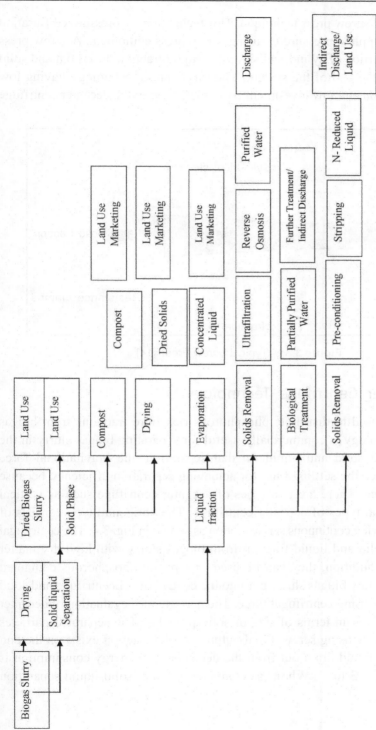

Figure 3.2: Overview of Biogas Slurry Processing Techniques

shows the setup of screw press technique. This technique is a pressurized filtration technology with applied pressure by using screw press equipment. A screw press fibre on the top portion of cylindrical sieve: the liquid fraction drains out and solid fraction exists at the end of the system. The screw press technology having low energy consumption and low investment cost as compare to the decanter centrifuge [10, 14].

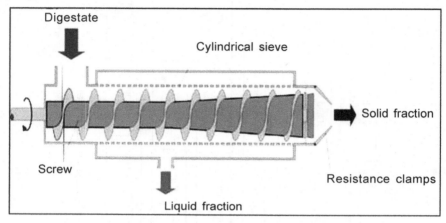

Figure 3.3: A Typical Screw Press [14]

4.2.2 Decanter Centrifuge Technology

This technology used to distribute phosphorous rich fibre fractions from biogas slurry. This technology is commercially useful for separating biogas slurry in the form of separated solid and liquid. In that separation, the gravitational force increase and reduce the settling time for achieving separation efficiency because of centrifugal force. There are two types of decanter centrifuge such as vertical type and horizontal types of decanter centrifuge. The horizontal type consists of closed cylinder having continuous screw motion as shown in Fig. 3.4. The centrifugal force separates solid and liquid fraction from biogas slurry with high dry matter concentration. In addition, they can be used to separate phosphorous contained biogas slurry. The raw biogas slurry put into the centre part of centrifuge and solid phase separated by using centrifugal force. The performance evaluation of decanter centrifuge carried out in terms of size of solid particles. The separated particles again compressed by using screw. Final output of solid fraction exists on the end of centrifuge and liquid drain out from the decanter. The energy consumption is high in the range of 3 to 5 kWh/m^3 as compared to other solid-liquid separation technologies [10].

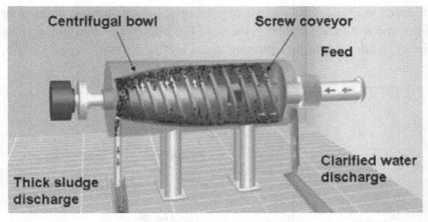

Figure 3.4: Decanter Centrifuge [13]

4.2.3 Belt Filter

For biogas slurry processing, the belt filters commonly used. Belt filters are divided into two categories: i) belt filter press and ii) vacuum belt filters. Belt filter press consists of a closed loop of textile belt, which is rounded around the cylinders as shown in Fig. 3.5.

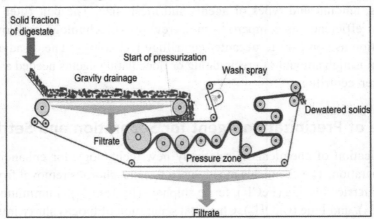

Figure 3.5: Belt Filter Press [10]

Raw biogas slurry is put continuously at the starting portion of the belt. The pre-dewatering process has done by centre of gravity at the time of feeding. In next stage, raw material is pressed between two belt filters. In that process, mechanical force is applied and cake is dewatered from outlet zone by using mechanical unit. The belt filter is washed by using washing spray. The washing material again used for the filtration. In vacuum belt filter, raw biogas slurry is fed into the filter belt and vacuum application carried out on underside of filter, which is illustrated in Fig. 3.6. The moisture content of biogas slurry is drained through the filter and solid cake stable on the belt. For improving solid liquid separation efficiency, adding some precipitating agent and flocculating agent in belt filter processing.

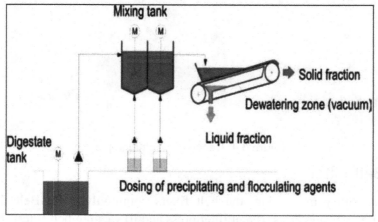

Figure 3.6: Vacuum Belt Filter [10]

The performance of biogas slurry separation are depends on biogas slurry properties, amount and types of agents and mesh size. The belt filter is higher separation efficiency as compare to the screw press technology and low energy consumption as compare to decanter centrifuge technology. The disadvantage of the system is high amount of precipitating or flocculating agents needed as compare to decanter centrifuge.

4.3 Use of Precipitation Agent for Separation and Settlement

The application of chemicals is relatively new technology for enhancing solid-liquid separation. The chemicals mostly used for phosphorous removal from biogas slurry are ferric chloride ($FeCl_3$), ferric sulphate ($Fe_2(SO_4)_3$), aluminium sulphate ($Al_2(SO_4)_3$), and lime ($CaOH_2$). Chemical separation of biogas slurry involves the adding chemicals to facilitate removal.

4.3.1 Processing of the Separated Solid Fraction

After separation, the obtained total dry concentration is 20-30%. This solid phase is stored and direct applied as bio fertilizer on the field. The separated solid fraction is mostly used as soil improver on agricultural land. After separation solids and it is desired to obtained marketable products, further processing is necessary in terms of composting and drying.

a. **Composting:** In that process, separated solid fraction mixed with organic fibrous material like woodchips. Composting is an ideal process in which microbes degrades and transfer organic matter under aerobic conditions. The produced composting mostly is used as bio-fertilizers and good performance as soil improver. The composting is applicable for vermiculture using earthworms.

b. **Drying:** Drying of solid fractions is carried out by using modern techniques such as belt dryer, drum dryer and solar dryers respectively. The dried solid fraction can be used as fertilizers for horticultural crops and gardening sectors. The modern utilization of solid fraction is pelletization of solid material. The solid pellets mostly used as fuel for cookstoves in rural areas. The dried solid fraction can be used also in nurseries or mushroom production business.

4.3.2 Processing of the Separated Liquid Fraction

Under separation process, the liquid fraction contains considerable amount of nutrients. Some part of liquid fraction is used for anaerobic digestion process. Liquid phase is applied on the heaps of compost. In that process, the use of liquid reduces the ammonia concentration and ammonia emission. There is several technologies are used for purification of liquid fraction such as nitrogen recovery by using ammonia stripping, nutrient concentration and water purification by using membrane technologies and evaporation.

5. ECONOMICS OF BIOGAS SLURRY PROCESSING

From economic point of view, the biogas slurry processing technologies is dependent on biogas slurry transportation cost, land application cost and production cost. It is very clear that the biogas slurry transportation cost and land use costs increases with increasing distances from the site. The processing cost will vary depending on the local conditions, transport distance and availability of raw materials.

6. CONCLUSIONS

Biogas plants are considered as an excellent way for energy generation. The slurry produced from the anaerobic digestion of organic residues, animal excreta and energy crop residues. Biogas slurry is an excellent plant fertilizer and improves nutrients management by processing. In many countries, biogas slurry management play vital role for environmental regulations. Biogas slurry processing is partial process, usually targeting volume reduction and solid-liquid separation for its utilization. In this partial process used relatively simple technologies for solid and liquid separation such as screw press technology and decanter centrifuge technology. These technologies are considered comparatively inexpensive and higher energy consumption for improving nutrient management in agriculture.

REFERENCES

1. Mathur AN, Rathore NS. Biogas production and utilization. Himanshu Publications, Udaipur. 1992.
2. Kadam R, Panwar NL. Recent advancement in biogas enrichment and its applications. Renewable and Sustainable Energy Reviews. 2017; ▨▨▨892-903. http:// dx.doi.org/10.1016/j.rser.2017.01.167.
3. Comparetti A, Febo P, Greco C, Orlando S. Current state and future of biogas and biogas slurry production. Bulgarian Journal of Agricultural Science. 2013; 19(1):1-14.
4. Balat M, Balat H. Biogas as a renewable energy source—a review. Energy Sources, Part A. 2009; 31(14): 12801293.http://dx.doi.org/10.1080/15567030 802089565.
5. Sujuta M, Prakash L, Kuwar TU. Global warming mitigation potential of biogas technology in security institution of Kathmandu Valley, Central Nepal. International Research Journal of Environment Sciences. 2014; 3(10): 68.
6. Ministry of New and Renewable Energy. Potential of renewable energy in India. (https://mnre.gov.in/). 2019; [accessed: 7 August 2019].
7. Dahiya AK,Vasudevan P, Biogas plant slurry as an alternative to chemical fertilizers. Biomass. 1986; 9(1): 67-74.
8. Islam R, Syed ME, Rahman MD, Rahman M, Deog Hwan OH, Chang RA. The effects of biogas slurry on the production and quality of maize fodder. Turk Journal of Agriculture. 2010; 34: 91-99.
9. Yu FB, Luo XP, Song CF, Zhang MX, Shan SD. Concentrated biogas slurry enhanced soil fertility and tomato quality. Acta Agriculturae Scandinavica Section B Soil and Plant Science. 2010; 60(3): 262-268.
10. Drosg B, Fuchs W, Al-Seadi T, Madsen M, Linke B. Nutrient recovery by biogas biogas slurry processing. IEA Bioenergy. 2015; 711.
11. Koszel M, Lorencowicz E. Agricultural use of biogas slurry as a replacement fertilizers. Agriculture and Agricultural Science Procedia. 2015; 7: 119-124. https://doi.org/10.1016/j.aaspro.2015.12.004.

12. Fangueiro D, Lopes C, Surgy S, Vasconcelos E. Effect of the pig slurry separation techniques on the characteristics and potential availability of N to plants in the resulting liquid and solid fractions. Biosystems Engineering. 2012; 113(2): 187-194. http://dx.doi.org/10.1016/j.biosystemseng.2012.07.006.

13. Hjorth M, Christensen KV, Christensen ML, Sommer SG. Solid–liquid separation of animal slurry in theory and practice. In Sustainable Agriculture. 2019; 2: 953-986. http://dx.doi.org/10.1051/agro/2009010.

14. Al Seadi T, Drosg B, Fuchs W, Rutz D, Janssen R. Biogas biogas slurry quality and utilization. In The Biogas Handbook. 2013; 267-301.

15. Groot LD, Bogdanski A. Bioslurry: brown gold. A review of scientific literature on the co-product of biogas production. Food and Agriculture Organization of the United Nations (FAO), 2013.

CHAPTER - 4

TECHNOLOGICAL UP-GRADATION IN BIOGAS PRODUCTION AND UTILIZATION FOR ENERGY GENERATION

Surendra R. Kalbande* and Sejal R. Sedani

*Department of Unconventional Energy Sources and Electrical Engineering
Dr. Punjabrao Deshmukh KrishiVidyapeeth, Akola-444104, Maharashtra, India
Corresponding Author

Nowadays greenhouse gas emission is increasing rapidly due to use of fossil fuel and industrialization results in climate change. To mitigate the various problems related to climate change, there should be advancement in renewable energy technology to cope today's need of energy with maintaining sustainability in the environment. Biogas is a multifaceted energy source having various applications such as heating, cooking, lighting, electricity and if biogas is purified, it can be used as a source of fuel in vehicles. This chapter deals with recent advances and technological up-gradation in biogas technology along with previous research work, which will give comprehensive synopsis of knowledge gleaned. it also annotated regarding the digester design, contemplation of anaerobic digestion, strategies of pretreatment, phase separation and codigestion; enzymatic hydrolysis, microbial stairs for enhancing biogas production, cryogenic separation, different biological and biogas reforming technologies. Elimination of contaminants and utilization of various energy crops for increasing the efficiency of an anaerobic digester is also discussed in this chapter. An eclectic amelioration in technical, social policies, economic and technology dissemination is obligatory to execute a sustainable energy system in the light of biogas as hereafter renewable source of energy.

1. INTRODUCTION

During the last decades, the insecurity of energy policies derived from the shortage of fossil fuels within the returning centuries, similarly because International concern for human-induced global climate change, have led to a part of the researching attention and resources have been target-hunting towards innovation and progressively growing use of renewable origin energies [1]. Many researchers have tried to develop strategies approaching to lower carbon dioxide (CO_2) emissions [2,8]. Within the various quite renewable energies, the utilization of biogas as associate energy supply is wide extended at associate industrial level because of the pliability offered by the method from that it comes – anaerobic digestion [9]. Biogas can be defined as the product of the anaerobic digestion biomass, consisting mainly of methane (CH_4) and CO_2 [10]. Biomass decomposition techniques to speed up biogas production has been studied like steam explosion for biomass decomposition through waste heat [11], sonification [12] or phytomass decomposition through water plasma expansion [13,14].

Biogas has lower emission rates than fossil fuel or the other fuel thus possesses abundant less environmental pollution potential compared to fossil fuels as shown in Table 4.1 [15]. Recent advancements in biogas technology have led to the development of more efficient AD systems incorporating such modifications as feedstock pre-treatment techniques, techno-economic gas upgrading, bioprocess improvements and advanced digester technologies among others.

Table 4.1: Comparison of Gaseous Emissions from Heavy Vehicles

g/kg	CO	HC	NO_X	CO_2	Particulates
Diesel	0.20	0.40	9.73	1053	0.100
Natural Gas	0.40	0.60	1.10	524	0.022
Biogas	0.08	0.35	5.44	223	0.015

2. BASIC PROCESS OF ANAEROBIC DIGESTION

2.1 Hydrolysis

During this reaction of the polymerized, mostly insoluble organic compounds, like carbohydrates and proteins, fats are converted into soluble monomers and dimers, for eg. monosaccharides, amino acids, and fatty acids. This stage of the methanogenesis passes through exoenzyme from the cluster of hydrolases (amylases, proteases, and lipases) created by acceptable strains of hydrolytic microorganism. Hydrolysis of complex polymers, that is, polysaccharide and cellulocottons is taken into account to be a stage that limits the speed of wastes digestion. During solid wastes digestion, solely 50 percent of organic compounds undergo biodegradation. The remaining a part of the compounds remains in their primary state owing to the shortage of enzymes collaborating in their degradation [16,17]. The rate of reaction method depends on parameters like size of particles, pH, and production of enzymes, diffusion, and surface assimilation of enzymes on the particles of wastes subjected to the digestion method. Hydrolysis is carried out by microorganism from the cluster of relative anaerobes of genera like eubacterium and Enterobacterium [18,19].

2.2 Acidogenesis (Acidification Phase)

During this stage, the acidifying bacterium converts soluble chemical substances, as well as hydrolysis product, to short-chain organic acids, alcohols, aldehydes, CO_2, and H_2. From decomposition of proteins, amino acids and peptides arise, which can be a supply of energy for anaerobic microorganisms. Acidogenesis could also be two-directional because of the results of varied populations of microorganisms. This method could also be divided into 2 types: hydrogenation and dehydrogenation. The essential process of transformations passes through acetates, CO_2 and H_2, whereas alternative acidogenesis product plays an insignificant role. Hence methanogens may directly use the new products as substrates and energy source. Accumulation of electrons by compounds like lactate, ethanol, propionate, butyrate, and better volatile fatty acids is that the microorganism response to a rise in H_2 concentration within the solution.

The new product might not be used directly by methanogenic microorganism and should be regenerate by obligatory microorganism manufacturing H_2 within the method known as acetogenesis. Among the product of acidogenesis, ammonia and H_2S that offer an intense unpleasant smell to the current section of the method ought to even be mentioned [16,20,21]. The acid phase bacterium belonging to facultative anaerobes use O_2 accidentally introduced into the method, making favorable conditions for the development of obligatory anaerobes of the subsequent genera: Pseudomonas, Bacillus, clostridia, Micrococcus, or Flavobacterium.

2.3 Acetogenesis

In this method, the acetate bacterium together with those of the genera of Syntrophomonas and Syntrophobacter convert the acid phase product into acetates and H_2 which can be employed by methanogenic bacterium [22]. Methanobacterium suboxydans account for disintegration of valeric acid to propanoic acid, whereas

Methanobacteriumpropionicum accounts for decomposition of propanoic acid to ethanoic acid. As results of acetogenesis, H_2 is discharged, that exhibits harmful effects on the microorganisms that perform this method. Therefore, interdependency is important for acetogenic bacterium with autophytic methane bacterium victimisation H_2, hereafter stated as syntrophy [22,23].

Acetogenesis may be a phase that depicts the potency of biogas production, as a result of around 70 percent of methane arises within the method of acetates reduction. Consequently, acetates are a key intermediate product of the method of methane digestion. In acetogenesis section, about 25 percent of acetates are made and about 11 percent of H_2 is created within the wastes degradation method [22].

2.4 Methanogenesis

This section consists within the production of methane by methanogenic bacterium. Methane during this section of the method is made from substrates that are the product of previous phases, that is, ethanoic acid, H_2, CO_2, and formate and methanol, methylamine, or dimethyl sulphide. Despite the very fact that solely few bacterium are able to produce methane from ethanoic acid, a huge majority of CH_4 arising within the methane digestion method results from ethanoic acid conversions by heterotrophic methane bacterium [24]. Only 30 percent of methane created during this method comes from carbon dioxide reduction carried out by autotrophic methane microorganism. During this method H_2 is employed up, that creates good conditions for the development of acid bacterium that bring about to short-chain organic acids in acidification phase and consequently to too low production of H_2 in acetogenic phase. A consequence of such conversions may be gas rich in CO_2, because only its insignificant part will be converted into methane [25,26].

3. LIMITATIONS OF SINGLE-STAGE DIGESTERS

Four guilds of microbes, which consists hydrolytic acidogens, non-hydrolytic acidogens, syntrophic acetogens, and methanogens, drive biomethanation method in a sequent and combined manner. Each of those guilds needs totally different conditions for best growth and performance [27]: Hydrolytic acidogens need massive surface areas to colonize and hydrolyse insoluble feedstock. Both hydrolytic and non-hydrolytic acidogens are additional active at comparatively low (5.5–6.0) hydrogen ion concentration [28].

Syntrophic acetogens and methanogens need stable neutral pH and are sensitive to even low concentrations of inhibitors (e.g., NH_3, H_2S, and SCFA). Syntrophic acetogens conjointly need close proximity with methanogens for economical interspecies H_2 transfer. However, most complete biomethanation systems in use are single-stage mesophilic digesters, within which it's troublesome to produce optimum conditions for all of the four guilds of microbes. As such, the metabolic activities of the microorganism guilds are compromised and also the performance of single-stage mesophilic digesters is commonly suboptimal; reduction of VS is quite slow and solely a little of VS will be converted, particularly once lignocellulosic biomass or microorganism biomass-laden biosolids are digestible [29].

Although pretreatments using heat and diluted acid or base will improve the digestibility of the feedstock, they inevitably increase capital and operational prices and probably turn out restrictive compounds [30]. In addition, up to 2/3 of the methane is created from acetate in anaerobic digesters [27], however syntrophic acetogens and acetoclasticmethanogens have extraordinarily slow growth because of their thermodynamically unfavorable pathways [31]. Consequently, the whole biomethanation method in single-stage mesophilic AD systems is commonly suboptimal and susceptible to being noncontinuous by accumulation of propionate and butyrate, particularly at high organic loading rates [32].

4. TEMPERATURE PHASED ANAEROBIC DIGESTION (AD) AS AN ALTERNATIVE TECHNOLOGY TO IMPROVE BIOMETHANATION

Thermophilic anaerobic digestion is taken into account one in all the foremost promising approaches to enhance biomethanation by fast hydrolysis of compound feedstock and different metabolism [33,34]. For lignocellulose-rich feedstocks, high temperatures enhance hydrolysis in all probability by loosening up the structure of lignocellulose in order that lignocellulose is more accessible to colonization by cellulolytic and xylanolytic microbes and their glycosyl hydrolases and by accelerating the enzymatic hydrolytic reaction. For microbicbiomass-laden feedstocks, high temperatures facilitate to lyse the intact microbic cells, creating the cellular parts available for bioconversion. However, many studies showed that thermophilic digester suffered from poor stability because of accumulation of SCFA, particularly propionate, reduced methane production, and enlarged CO_2 content [32, 35]. The above limitations related to thermophilic AD are thought to be owing to many factors. First, elevated temperatures decrease diversity and strength of methanogens in digesters as solely 3 species of methanogens are known in thermophilic anaerobic digesters [36,37]. Second, extreme temperature decreases the solubility of H_2.

Third, some microbes, particularly syntrophic acetogens and methanogens, are more at risk of repressive metabolites (e.g., NH_3, H_2S, and propionic and butyric acids) at thermophilic temperatures than at mesophilic temperatures [27]. Efficiency and stability/reliability of biomethanation processes are crucial for economically viable production of methane biogas as energy. However, TPAD remains in its infancy and principally operated through empirical observation at laboratory or pilot scales. Further analysis is required to support its wide implementation at industrial scales. To this finish, laboratory and pilot scale studies on TPAD systems, together with analysis of energy and economic balance, ought to be conducted on the categories of feedstocks to be digestible to supply enough information for design, implementation, and operation of large-scale TPAD systems. A thorough understanding of the microbe in anaerobic digesters will eventually help with rationale design, modeling, and operation of TPAD systems.

5. COMMUNITY SIZED BIOGAS PLANT FOR ELECTIRICITY GENERATION

5.1 Main Components of Biogas Plant

The main components of biogas plant are inlet tank, digester, dome, outlet chamber. The dimensions of 50 m^3 modified biogas plant are shown in Table 4.2.

Table 4.2: Dimensions of 50 m^3 Modified Biogas Plant

S.N.	Part Name	Symbol	Dimension (cm)
1.	Diameter of digester	D	442
2.	Inner radius of digester	R	221
3.	Depth of digester	H	442
4.	Height of outlet opening	H1	217.5
5.	Height of smaller portion of outlet chamber	H2	360
6.	Length of bigger portion of outlet chamber	M	540
7.	Width of bigger portion of outlet chamber	N	390
8.	Diameter of mixing tank	R	240
9.	Height of mixing tank	P	60

Technological Up-gradation in Biogas Production and Utilization for Energy Generation

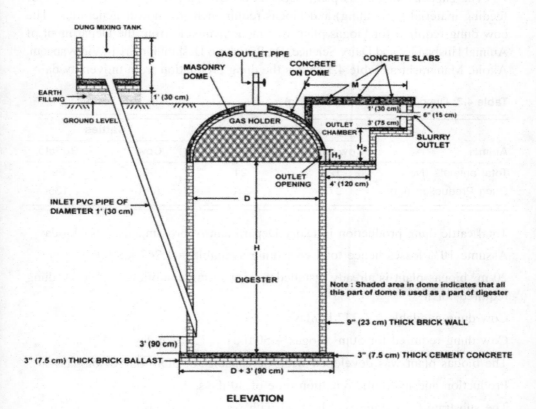

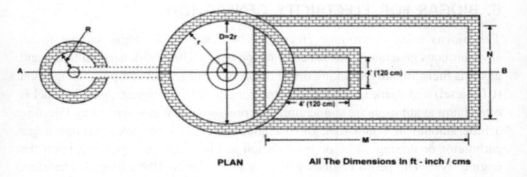

Figure 4.1: Schematic of Fifty Cubic Meter Modified Biogas Plant

The capacity of biogas plant depends on two parameter i.e. Availability of feeding material (cow dung) and biogas requirement for power generation. The cow dung required for biogas plant was made available from the Department of Animal Husbandry and Dairy Science, Dr. Panjabrao Deshmukh Krishi Vidhyapeeth, Akola, Maharashtra. Table 4.3 shows the dung production from university dairy.

Table 4.3: Dung Production per Day (Kg)

Animal	Calves		Cattles	
	Cow calves	Buffalo calves	Cow	Buffalo
Total animals (No.)	122	21	62	09
Dung Production (kg)	976	210	620	135

Total cattle dung production in Dairy Department Approximately 1941 kg/day

Assume 10% losses hence total cow dung available = 1,747 kg/day

20 m^3 biogas plant is already installed & in working condition, so 500 kg dung required for it.

Cow dung available = 1,247 kg/day

Cow dung required for 50m^3 biogas = 50/0.04 = 1250 kg/day

The biogas plant was developed with the capacity of 50m^3/day of biogas.

Production and hydraulic retention time of 30 days.

The substrate Cow dung will be feed with water in 1:1 ratio.

6. BIOGAS FOR ELECTRICITY GENERATION

The biogas used to produce electricity by coupling a biogas engine to an asynchronous generator. Internal combustion engines that work each on liquid and gaseous fuels, broadly speaking divided into two categories: compression ignition (CI/ diesel) and spark ignition (SI/ petrol) engines. In CI engines, gasified fuel is often burnt at atmospheric temperature and pressure and fuel is ignited by injecting a little amount of liquid fuel. In SI engines, the gas is inducted through a gas carburetor or mixing value or by injection at low pressure depending upon the engine style. The petrol engines will be used to run on 100% biogas. Therefore large capacity biogas plants are to provide to produce gas for running IC engines (Fig. 4.2).

Figure 4.2: Biogas Power Generation Unit

The traditional consumption rate of biogas for running IC engines is 0.45 to 0.54 m^3/h/HP or 0.60 to 0.70 m^3/h kW if used for power generation. One cubic meter of biogas generates 1.25 kWh electricity which might run one HP motor for two hour.

7. PERFORMANCE OF BIOGAS-BASED GAS ENGINE

The engine performance was evaluated on no load, half load and full load condition and it was run fully on 100% biogas having specifications depicted in Table 4.4.

Table 4.4: Specification of Biogas Engine

S.N.	Particulars	Specification
1.	Type	Vertical, Water Cooled
2.	No. of cylinder	Four
3.	Rated bhp (with biogas)	6 kW
4.	Rated RPM	1500
5.	Fuel	Biogas

A three phase 7.5 kVA AC generator of frequency 50 Hz, Voltage 230/415 volts and speed 1500 rpm was coupled to the engine through a torque Transducer to produce electricity. The electrical output was measured by recording the voltages across the three phases of the AC generator and the current drown by the resistive loading system. The voltage, current, frequency and power output were measured by multimeter and average values were considered for calculation.

Brake power

$$\text{BSFCE} = \frac{\text{Biogas flow rate (m}^3\text{/h)}}{\text{Brake Power (kW)}} \quad \ldots (1)$$

Brake specific fuel consumption of engine (BSFC$_E$)

$$\text{BSFCE} = \frac{\text{Biogas flow rate (m}^3\text{/h)}}{\text{Brake Power (kW)}} \quad \ldots (2)$$

Brake specific energy consumption (BSEC)
BSEC = Calorific value of gas × BSFC

Brake thermal efficiency
BTE (%) = 360/BSEC ... (3)

8. ECONOMIC ANALYSIS OF THE BIOGAS POWER GENERATION SYSTEM

The following assumptions were taken into consideration for carrying out economic assessment of the biogas power generation unit.

1. The 50m^3 capacity of the biogas plant will require 1250 kg/day cow dung.
2. The fixed cost of biogas plant and power generation unit was Rs. 5,84,000/- and Rs. 4,84,313/- respectively and the life of the unit is considered to be 20 years.
3. The fresh cattle dung 37,500 kg/month (1250 kg/day) is required having total solid content of 10% produced 3750 kg/month dried bioslurry.
4. The cost of cattle dung per kg is Rs. 0.5 hence total cattle dung cost is Rs. 18750/month.
5. An Average biogas production per month was 27.24 m^3/day and for 1 h operation 4 m^3 biogas was required for running engine 6.5 h/day.
6. Assume 80% engine efficiency of 6 kW thus electricity production per day would be 31.3 kW (936 kW/month).
7. The average selling price of enrich digested slurry was Rs. 10/- and price of electricity was Rs. 10/kWh.
8. The cost of labour was considered as Rs. 220/day.

The annual repair, maintenance and replacement cost was considered to be 10% of the initial investment. Table 4.5 shows the capital statement and economic of the power generation from biogas system.

Table 4.5: Capital Statement and Economic of the Power Generation from Biogas System

S.N.	Operational parameter	Cost (Rs)
1.	Cattle dung required (kg)	1250
2.	Engine operation time (h/day)	6
3.	Electricity produced (kWh/day)	31.2
4.	Nutrient enrich dried bio-slurry (kg/month)	3750
A	Installed capital cost of system, (Rs.)	1072913
B	Operational cost of system, (Rs.)	
	i) Cost of cattle dung Rs/Month	18750
	ii) Labour cost, Rs/month (@ 2 labor /day)	13200
	iii) Annual maintenance cost, 1% on installed cost	894
	Total operational cost Rs/ month	**32844**
C	**Net gain due to system application, Rs/ month**	**46860**

The techno-economic feasibility of the biogas power generation unit were computed by considering the initial investment of the biogas plant and power generating unit, average repair and maintenance cost, cost of raw material and selling price of the digested slurry after drying. The average parameter was drawn on the basis of experimental results. Economics of biogas power generation unit, the different economic parameter of Biogas power generation unit is summarized in Table 4.6.

Table 4.6: Economic Analysis of Biogas Power Generation Unit

S.N.	Description	Biogas power generation unit
1.	Initial investment (Rs)	1072913
2.	Operation and maintenance cost (Rs/month)	32844
3.	Net gain due to system application, Rs/month	46860
4.	B:C ratio	1.86
5.	Payback period	3 years
6.	NPW (Rs.)	2217545.32

9. EVALUATION OF PERFORMANCE OF COMMUNITY SIZED 50 m³ BIOGAS PLANT

9.1 Characterization of Cattle Dung and Digested Slurry

The average moisture content in fresh cow dung were 79.70 percent and after digestion 89.12 percent during on field experiment. The average total solids in fresh cow dung were 20.31 percent and in digested slurry 11.02 percent. The average volatile matter in fresh cattle dung 20.31 percent and in digested slurry were 11.05. It was found that Carbon, Nitrogen and Oxygen content more in digested slurry than the fresh cow dung. But Hydrogen content was less found than fresh cow dung as given in Table 4.7.

Table 4.7: Characterization of Cattle Dung and Digested Slurry

Proximate Analysis					
Samples	Moisture Content (%)	Total Solid (%)	Volatile Matter (%)	Ash Content (%)	Fixed Carbon (%)
Cattle dung	79.79	20.31	15.83	3.60	1.06
Digested slurry	89.12	11.05	7.83	2.19	1.03
Ultimate Analysis					
Samples	Carbon (%)	Hydrogen (%)	Nitrogen (%)	Oxygen (%)	
Cattle dung	27.51	13.44	1.78	53.78	
Digested slurry	32.09	6.82	1.94	56.95	

10. ANALYSIS OF NUTRIENTS IN COW DUNG AND DIGESTED SLURRY

The cattle dung and digested slurry was analyzed for nutrients like nitrogen, phosphorus and potassium at every 7 days interval. Significant differences in N, P and K content of cattle dung before and after digestion were observed. In cattle dun N, P and K Observed average 1.13, 0.75 and 0.55 respectively. And in digested slurry was 1.33, 0.90 and 0.68 respectively. The observations are represented in Table 4.8.

Table 4.8: NPK Content in Cow and Digested Slurry

S. N.	Particular	Cow Dung	Digested Slurry
1.	N (%)	1.13	1.33
2.	P (%)	0.75	0.90
3.	K (%)	0.55	0.68

11. BIOGAS PRODUCTION

The biogas production for the 12th of June to 12th July 2019 has been shown in Fig. 4.3. Average ambient minimum and maximum temperature was 27.12 and

35.15°C. Biogas production from fixed dome biogas plant varied from 22 to 40 m³/day. Average biogas production from fixed dome biogas plant was 27.24 m³/day.

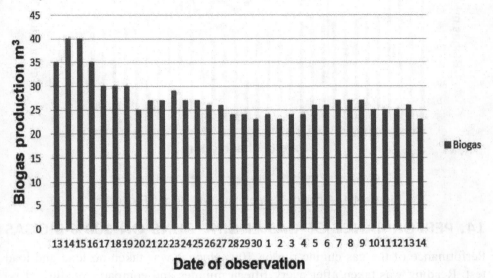

Figure 4.3: Biogas Production in Month of June and July

12. EFFECT OF AMBIENT TEMPERATURE ON BIOGAS PRODUCTION

Ambient temperature is directly effect on biogas production, during study it was observed that biogas production decrease according to ambient temperature decreased. Observation was taken in months of June and July. During study period found that maximum and minimum temperature were 41 and 23°C respectively. The maximum and minimum biogas production was observed 40 and 22 m³/day respectively and average biogas production as per Fig. 4.4.

13. COMPOSITION OF BIOGAS

The average methane content in biogas production of 5 weeks was 57.51 percent and carbon dioxide was 41.33 percent. For the first 10 days the methane content was around 54 to 57 percent, thereafter it was above 57 percent for the remaining period. The maximum methane content in the biogas production was 57.43 percent.

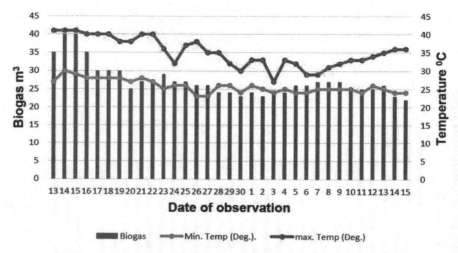

Figure 4.4: Ambient Temperature and Biogas Production

14. PERFORMANCE OF GAS ENGINE RUNS ON 100% BIOGAS

Performance of the gas engine run on 100% biogas was taken no load and load test. Reading was taken after every fifteen minutes and compares reading of no load and load test. Found that average biogas consumption (m^3), Speed of engine (rpm), Output voltage (Volt) and current (Ampere) of the no load test were 0.79, 1394, 313 and no current shown and on load test 1.00, 1299, 299 and 6. On no load test fuel consumption was less however rpm more than the load test. The engine run stable on 100% biogas and might produce electricity. The utmost rpm (revolution per minutes) which will be obtained by victimization 100% biogas was found regarding 1500 rpm [38,39].

15. BIOGAS-ENHANCEMENT STRATEGY

The pretreatment hydrolysis step is that the rate-limiting issue for the whole AD method [40]. However, initial substrates are generally terribly complicated in nature; pretreatment therefore solves this issue by fixing the composition of substrate so as to be additional appropriate for digestion [41]. Pretreatment ways consist of physical/mechanical (e.g., ultrasound, high, analysis), electrical pulses, wet chemical reaction, freeze/thaw, chemical (e.g., ozonation, alkali, acids), biological and thermal (<100°C, >100°C, microwave) techniques. Particle size reduction will increase the extent, providing a higher contact between anaerobic micro-organism and also the substrate [42]. Mechanical pretreatments (e.g., sonication, lysis-centrifugation, collision, liquid shear, high pressure homogenization, phase change, and maceration) are employed in order to scale back the substrate's particle size

[43]. These pretreatment ways were largely applied at laboratory scale, with solely a couple of ways with success used at massive scale. Only a couple of samples of the thermal hydrolysis method like Cambi, Porteous, and Zimpro processes, and thermo chemical pretreatment strategies like Synox, Protox, and Krepro are applied on an outsized scale [44]. Bacteria will produce 2 kinds of enzymes: endoenzymes (i.e., within the cell) and exoenzymes (i.e. secreted outside the cell) within the case of AD, the first step is hydrolysis of substrate. Bacteria will produce two kinds of enzymes: endoenzymes and exoenzymes. Within the case of AD, the first step is hydrolysis of substrate. It is doable to separate the biological pretreatment of biomass/substrate before it progresses to any AD stages like acidogenesis, acetogenesis, and methanogenesis. It will either be treated with totally different extracted enzymes or microbic strains that produce hydrolytic enzymes. The hydrolytic enzymes are in the main cellulase, cellobiase, amylase, xylanase, protease, and lipase. No bacteria produces all of the exoenzymes required for the degradation of massive sort of substrates; thus large and various communities of bacterium that produce completely different forms of enzymes are needed. Many authors have reported the utilization of crude and commercial enzymes so as to treat the complicated organic matters of biomass for improvement of biogas production [45]. Many studies are according on the hydrolysis of lignocellulose and polysaccharide (i.e., primary constituents of plant materials) by lingo-cellulase and cellulase, severally [46]. Similarly, improvement of biogas was also according within the literature by using lipase throughout the pretreatment of lipid-rich wastewaters [45]. Production of enzyme for the hydrolysis step through genetically designed microorganisms is also useful for biogas improvement. The engineered yeast Saccharomyces cerevisiae has been reported to ferment cellulose, xylose, and arabinose [46].

Cellulolytic prokaryotes (e.g. Actinomycetes and alternative mixed consortia) were according to boost biogas production from the oxen dung [47]. It was found that a-c-d coculture addition (3 %-9 % as an inoculum) to feces fermentation was helpful in increasing gas and methane production by 56.36 % and 18.09 %, respectively, when compared to that of natural fermentation by feces microbes [48]. Microorganisms present within the rumen have the power to anaerobically degrade cellulosic waste. Many researchers have tried to isolate completely different cellulolytic microorganisms that possess the next degradable activity. Sarkar et al. [49] reported eleven totally different microbic consortia that had different enzymatic activity and reduced the time span for degradation for organic food waste. Dhadse et al. [50] studied biogas production through 3 totally different microorganism consortia "A," "B," and "C." Where consortia "B" contains strict (obligate) and facultative anaerobic microorganism (Bacteroides, Peptostreptococcus, eubacteria, and Propionibacterium); consortia "C" represents methanogenic microorganism (Methanobacterium formicicum, Methanobrevibacterruminantium, Methanisarcina Frisia, Methanothrix soehngenii) whereas consortia "A" represents all the eight isolates. Consortia "C" yielded the largest methane concentration i.e. 76% in comparison to consortia "A" of 23% and consortia "B" of 1%. The AD method is often separated into 2 phases: acidogenesis and methanogenesis. Main product of Acidogenesis, or acid fermentation is volatile fatty acid (VFA). Methanogenesis produces methane and CO_2 using the end product of the primary section. The main advantage of phase separation is that the toxicity of the acid created in one section isn't directly encountered with methanogens within the second section. Another advantage for a separate reactor is separate controlling of an environmental condition. The two phases are very different in their microbial populations, digestion rates, and environmental conditions. Different reactor arrangement that employ two-phase AD with two separate reactors include a) upflow anaerobic sludge blanket (UASB)— UASB system, b) continuous stirred tank reactor (CSTR)—upflow anaerobic filter system, c) hybrid reactor, d) CSTR—anaerobic fluidized bed reactor system, e) two-phase plug-flow reactor and f) anaerobic packed bed reactor [51]. The digestion of one substrate having a low carbon:nitrogen (C : N) quantitative relation or repressing compounds might cause low biogas production with lower quality. Low carbon results in low methane series production and high N_2 content results in ammonia gas or ammonia accumulation that is a restrictive methanogenic activity. In such a scenario, improvement of methane production is increased by the synchronous codigestion of the substrate with a high carbonic substrate (high C : N ratio) which might be optimized by employing a totally different mix ratio and kinds of cosubstrate as a result of it enhances the synergism, methane production, digestate quality, and dilutes restrictive compounds [51]. However, a better C : N ratio results in higher gas production however lowers buffer capability, whereas a low C : N ratio results in lower gas production and higher buffer capability [52, 53].

16. RECENT ADVANCES IN BIOGAS PURIFICATION TECHNOLOGIES

Biogas created through AD contains methane, CO_2, hydrogen, H_2S sulfides, vapor, ammonia, and siloxanes. Among them, solely methane and H_2 are often used for energy functions. Without purification of the biogas, contaminated gases adversely have an effect on the appliances (i.e., H_2S is corrosive to engine or pipeline). Hence, concentrating methods have been developed. Among the contaminated gases, CO_2 is taken into account to be the main gas whereas others are present in trace amounts. Water scrub works on the idea of H_2S and CO_2's higher liquid solubility compared to CH_4 [54]. Through this method, 80%-99% CH_4 purity can be obtained [55]. This accounts for 41 percent of the global biogas market upgrading because of enough accessibility of water at low price and is less sensitive to biogas impurities [56]. To support good gas-liquid mass transfer efficiency, countercurrent operation is used in which the absorption and desorption units have random packing of Pall or Raschig rings [57, 58]. Scrubbing can also be done by organic solvents such as PE glycol basedabsorbents (Selexol, Genosorb) with a higher affinity to CO_2 and H_2S than water. In this method, biogas and organic solvents are cooled down at 20°C before absorption [58]. The main advantage of this method is that the anticorrosion nature of solvents however shares solely 6 percent of the upgraded biogas market [56]. In order to avoid sulfur-mediated deterioration of the organic solvent, either the complete removal of H_2S using activated carbon filters before organic solvent scrubbing or regeneration of organic solvent using steam or inert gas after desorption process is recommended [57]. Organic solvents like tetradecane or Selexol will take away the siloxane at 97%-99% efficiencies [54]. Water absorption in glycols can even at the same time take away oil and dirt particles. Chemical cleaning is concerned within the carbon dioxide reactive absorbents like alkanol amines (monoethanolamine, diethanolamine, etc.) or alkali liquid solutions (NaOH, KOH, CaOH, K_2CO_3, etc.) [54,59]. Nowadays, a combination of methyldiethanolamine (MDEA) and piperazine constitutes major amine absorbents, that are wide used commercially at $MDEA:CO_2$ molar ratios of 4:7 [58]. In chemical cleaning, intermediate chemicals like $(CO_3)^2$, (HCO_3) are generated exothermally once the adsorbable CO_2 reacts chemically present within the scrubbing solution, leading to higher absorption capacities [57]. This process is reported to share 22% of global upgrading market [56]. The selective adsorption of CO_2 over CH_4 on porous adsorbents with a large surface area includes activated carbon, silica gel, activated alumina, zeolite, and polymeric sorbents [54, 57, 60]. Molecular size exclusion and surface assimilation affinity are the essential separation mechanism of pressure swing adsorption (PSA) technology. The average pore size of adsorbents is 3.7 Å, which retains CO_2 (3.4 Å), excludes CH_4 (3.8 Å), and makes up 21% of the global upgrading market [54, 56]. For siloxane removal of up to 95 percent, Zeolites, activated alumina, silica gel, and activated carbon adsorption supports when treating dry biomethane. Adsorption of water has been carried out in packed bed columns with silica, alumina, magnesium oxide, or activated carbon under pressure (6-10 bar) [54, 61]. In general, either polymeric materials or mixed matrix membranes (MMMs) structure membranes. Examples of polymeric materials embrace cellulose ester, cellulose acetate, polyimides (PI), polyetherimide (PEI), polyamide, PSF, PES, polycarbonate (brominated), polyphenylene oxide, polymethyl pentene, polydimethylsiloxane, and polyvinyltrimethylsilane [62]. MMMs are prepared by dispersing selective inorganic materials into endless polymer matrix to enhance performance. Three completely different forms of special fillers are used to prepare MMMs: (1) ordered mesoporous silicas (OMSs), (2) high aspect ratios (HARs) silica-based particles and (3) metal-organic frameworks (MOFs). OMSs have ordered mesoporosity of the surface that permits for penetration of the compound chain [63]. Some examples are MCM-41/(PSF/PEI), MCM-48/(PSF/PEI). Thin layer with a high area/volume aspect ratio is obtained by swelling and exfoliation of microporous material [63]. A multistage system using numerous membrane modules could be a sensible strategy to get high purity and high recovery of alkane series [64]. Four differing kinds of membrane modules are used in multistage systems by applying totally different membrane in an exceedingly cascade mode and recycle loop. In the hybrid processes, membrane-based biogas upgrading technology is combined with typical gas separation instrumentation and should be superior to singular membrane technology. In another approach by Makaruk et al. [65], permeate of the membrane separation contained CH_4 which will be used to drive power engine and at the same time heat generated in engine is used for heating the fermentation method. Different biogas parts have the power to liquefy or solidify at numerous low temperatures. This characteristic is used for the selective separation of biogas elements. In cryogenic separation at constant pressure (10 bar), the temperature of biogas was decreased stepwise to 225°C (i.e., where F_2O, H_2S, siloxanes, and halogens are removed in the liquid phase), 255°C (i.e., where most CO_2 is liquefied), and finally to 285°C (i.e., where the remaining CO_2 solidifies) [54]. Even once decreasing the temperature to 2162°C to 2182°C, liquefied biomethane might be generated [54]. Allowing this process to operate at high pressure (80 bar) prevented the sudden solidification of CO_2 below 278°C, avoided the clogging of pipelines, and heat exchanges [57, 58, 60]. This technology provided 97 percent pure biomethane with lower (2%) loss however can't be commercial and shares solely 0.4 percent of the worldwide market [56, 59]. It was suggested to get rid of water, H_2S, siloxanes, and halogens before carbon dioxide removal to avoid pipe or heat exchanger hindering [58]. Investment prices and energy inputs ought to be evaluated for additional operation of this technology. Carbon dioxide reduction by biological means that is representing its cell component is an eco-friendly approach for biogas up gradation. H_2-assisted carbon dioxide bioconversion, microalgae-based carbon dioxide fixation, enzymatic carbon dioxide dissolution, fermentative carbon dioxide reduction, and in place carbon dioxide desorption are biological strategies for the removal of carbon dioxide from biogas.

17. BIOGAS REFORMING TECHNOLOGIES

In stream reforming process, under the presence of a catalyst (e.g., transition metals: Pt, Ni; noble metals: Rh, Pd), methane reacts with water vapor to produce CO and H_2 at 650°C-850°C. CO can be separated from H_2 by a water gas shift reaction using a suitable catalyst (e.g., Cu, Fe, Mo, and Fe-Pd alloys) at 300°C-450°C [66, 67]. In this method, carbon deposited on the catalyst's surface could cause deactivation. It is solely helpful within the case of carbon deposited within the kind of nanotubes that preserves longer catalyst activity. The membrane filter-based reactors used for steam reforming (SR) permits all reactions to occur in a single vessel, including separation process. Lin et al. [68] prepared mesoporous $Ni_2xCe1-xO_2$ catalysts through a reverse precipitation method. By using this catalyst in a SR method (500°C-900°C), they obtained a better H_2:CO ratio and a lower H_2:CO_2 ratio than once a commercial catalyst was used. Increasing the nickel content within the catalyst raised methane conversion to H_2 production. Xu et al. [69] developed a polymer solution fuel cell to get high H_2 (70%) from desulfurized biogas. They utilized a steam reformer at high temperature, two water-gas shift reactors at low temperatures, a selective CO oxidizer, and a gas producer. Italiano et al. [70] invented a nano crystalline Ni/CeO_2 catalyst for biogas conversion to H_2 through oxy-steam reforming. They obtained above 85% H_2 yield at 800°C when this catalyst was used. Differing from the SR method, partial oxidization reforming (POR) is very energy-releasing in nature during which methane is part catalytically oxidized with O_2 into syngas (H_2:CO ratio is 2:1) at atmospheric pressure and 700°C-900°C so as to cut back coke formation [71]. Similar to the SR method, auto thermal reforming (ATR) is energy-absorbing in nature and combines SR and POR technologies. In ATR method, heat generated throughout the partial oxidisation of methane is used in parallel for the SR process. The ratio of syngas (H_2:CO) is between 2.0 and 3.5 using O_2:CH_4 and H_2O:CH_4 ratios between 0.25:0.55 and 1.0:2.5, respectively [72,73]. In the dry reforming (DR) method, CH_4 reacts with carbon dioxide to supply syngas (H_2, CO) at temperatures between 700°C and 900°C. A 1-1.5 ratio of CH_4 and carbon dioxide yields up to 50 percent H_2 [74]. The advantage of this method is that the utilization of 2 greenhouse gases, whereas disadvantages embrace its endoergic nature requiring external energy similarly as coke formation on the active surfaces of the catalyst. The catalysts used for this process are primarily Rh, Ru, Pt, Co, and Ni [75,76]. Dry reaction reforming is that the exo- and endothermic combination of DR and POR processes, employing a parallel feeding of O_2 with methane and carbon dioxide [71]. This technique controls the deposition of coke on the catalyst also as improves the CH_4 conversion to higher H_2 yield using lower energy input. The exoergic nature will be controlled to totally different degrees by the O_2 feed's concentration. Bimetallic catalysts are employed in this method to boost the soundness of the catalyst [77,78]. Purwanto and Akiyama [79] reported passing biogas (CH_4, CO_2) endlessly through a hot slag packed bed reactor at atmospheric pressure to supply syngas with higher H_2 yields. In this method, slag acted as the thermal media likewise the catalyst for biogas element decomposition. Increasing the temperature of the slag resulted in higher H_2 yields and higher methane conversions (96%). Chun et al. [80] studied plasmatron-assisted CH_4 reforming in which high temperature plasma was generated by air (5.1 L min 21) and arc discharge (6.4 kW). In the parametric study, they found the optimum methane flow ratio is 38.5%, methane conversion rate is 99.2%, syngas concentration i.e H_2 =45.4%; CO = 6.9%; CO_2=1.5%; and C_2H_2= 1.1%, H_2:CO ratio is 6.6, H_2 yield is 78.8% and reformer thermal potency is 63.6%. Rueangjitt et al. [81] investigated the upgrading of biogas by passing the biogas through a multistage AC (alternating current) gliding arc system. A fast increment within the stage numbers of plasma reactors, applied voltage, and electrode gap distance raised the conversion of CH_4 and carbon dioxide to H_2. The performance may be improved by employing a combination of plasma reforming and POR.

18. CONCLUSIONS

From the result and discussion it was observed that the nutritional value of the digested slurry had more than the fresh cow dung and it will give good return on investment, the digested slurry can use as fertilize or use it after drying. The fifty cubic meter biogas plant is feasible for the power generation and electricity used in agriculture provide trough the biogas power generation unit. From the economic analysis biogas power plant give return to the farmer.

19. ACKNOWLEDGMENT

The authors gratefully acknowledge to NAHEP-ICAR for providing funds for carrying out research work.

REFERENCES

1. Anneli P, Wellinger A. Biogas upgrading technologies–developments and innovations. IEA Bioenergy. 2009:20. doi:10.1016/j.wasman.2011.09.003.
2. Chen Q, Acey C, Lara JJ. Sustainable futures for linden village: A model for increasing social capital and the quality of life in an urban neighborhood. Sustainable Cities and Society. 2015; 14(1): 359–73. doi:10.1016/j.scs.2014.03.008.

3. Fatih DM, and Balat M. Progress and recent trends in biogas processing. International Journal of Green Energy. 2009; 6(2): 117–42. doi:10.1080/15435070902784830.
4. Lam CH, Sabyasachi Das NC, Erickson CD, Hyzer MC, Anderson JE, Wallington TJ, Tamor MA, Jackson JE, Saffron CM. Towards sustainable hydrocarbon fuels with biomass fast pyrolysis oil and electrocatalytic upgrading. Sustainable Energy & Fuels. 2017; 1(2): 258–66. doi:10.1039/C6SE00080K.
5. Sattary S, Thorpe D. Potential carbon emission reductions in Australian construction systems through bioclimatic principles. Sustainable Cities and Society. 2016; 23: 105–13. doi:10.1016/j.scs.2016.03.006.
6. Xie Y, Björkmalm J, Chunyan M, Willquist K, Yngvesson J, Wallberg O, Xiaoyan J. Techno-economic evaluation of biogas upgrading using ionic liquids in comparison with industrially used technology in Scandinavian anaerobic digestion plants. Applied Energy. 2017. doi:10.1016/j.apenergy.2017.07.067.
7. Yang Y, Ajmal S, Zheng X, Zhang L. Efficient nanomaterials for harvesting clean fuels from electrochemical and photoelectrochemical CO_2 reduction. Sustainable Energy and Fuels. 2017. doi:10.1039/C7SE00371D.
8. Zhao X, Pan W, Weisheng L. Business model innovation for delivering zero carbon buildings. Sustainable Cities and Society. 2016; 27: 253–62. doi:10.1016/j.scs .2016. 03.013.
9. Younis M, Alnouri SY, Abu Tarboush BJ, Ahmad MN. Renewable biofuel production from biomass: A review for biomass pelletization, characterization, and thermal conversion techniques. International Journal of Green Energy. 2018; 15(13): 837–63. Taylor & Francis. doi:10.1080/15435075.2018.1529581.
10. Khan U, Imran M, Othman HD, Hashim H, Takeshi Matsuura AF, Rezaei-Dasht Arzhandi IM, Wan Azelee I. Biogas as a renewable energy fuel – a review of biogas upgrading, utilization and storage. Energy Conversion and Management. 2017. doi:10.1016/j.enconman.2017.08.035.
11. Marousek J, Stehel V, Vochozka M, Marouskova A, Ladislav K. Postponing of the intracellular disintegration step improves efficiency of phytomass processing. Journal of Cleaner Production. 2018; 199: 173–76. doi:10.1016/j.jclepro.2018.07.183.
12. Jeon BH, Choi JA, Kim HC, Hwang JH, Abou-Shanab RAI, Dempsey BA, Regan JM, Kim JR. Ultrasonic disintegration of microalgal biomass and consequent improvement of bioaccessibility/bioavailability in microbial fermentation. Biotechnology for Biofuels. 2013; 6: 37. doi:10.1186/1754-6834-6-37.

13. Marousek J, Kwan JTH. Use of pressure manifestations following the water plasma expansion for phytomass disintegration. Water Science and Technology. 2013; 67: 1695–700. doi:10.2166/wst.2013.041.

14. Tian Y, Zhang H. Producing biogas from agricultural residues generated during phytoremediation process: Possibility, threshold, and challenges. International Journal of Green Energy. 2016; 13: 1556–63. doi:10.1080/15435075.2016.1206017.

15. Vijay VK, Chandra R, Subbarao PMV, Kapdi SS. Biogas purification and Bottling into CNG cylinders: producing Bio-CNG from biomass for rural automotive applications, The 2nd Joint International Conference on "Sustainable Energy and Environment (SEE 2006)", 2006.

16. Conrad R. Contribution of hydrogen to methane production and control of hydrogen concentrations in methanogenic soils and sediments, FEMS Microbiology Ecology. 1999; 28(3): 193–202.

17. Parawira W, Read JS, Mattiasson B, Bjornsson L. Energy production from agricultural residues: high methane yields in pilot-scale two-stage anaerobic digestion, Biomass and Bioenergy, 2008; 32(1): 44–50.

18. Bryant M. Microbial methane production: theoretical aspects. Journal of Animal Science. 1979; 48(1): 193–201.

19. Smith P. The microbial ecology of sludge methanogenesis. Developments in Industrial Microbiology. 1966; 7: 156–161.

20. Claassen PAM, Lopez Contreras AM, Sijtsma L. Utilisation of biomass for the supply of energy carriers.Applied Microbiology and Biotechnology. 1999; 52(6): 741–755.

21. Ntaikou G, Antonopoulou, Lyberatos G. Biohydrogen production from biomass and wastes via dark fermentation: a review. Waste and Biomass Valorization. 2010; 1(1): 21–39.

22. Schink B. Energetics of syntrophic cooperation in methanogenic degradation. Microbiology and Molecular Biology Reviews. 1997. 61(2): 262–280.

23. Bok FAM, Harmsen HJM, Plugge CM. The first true obligately syntrophic propionate-oxidizing bacterium, Pelotomaculum schinkiisp. nov., co-cultured with Methanospirillum hungatei, and emended description of the genus Pelotomaculum. International Journal of Systematic and Evolutionary Microbiology. 2005; 55(4): 1697–1703.

24. Verstraete W, Doulami F, Volcke E, Tavernier M, Nollet H, Roles J. The importance of anaerobic digestion for global environmental development. In Proceedings of the JSCE Annual Meeting 2002: 97–102.

25. Griffin ME, McMahon KD, Mackie RI, Raskin L. Methanogenic population dynamics during start-up of anaerobic digesters treating municipal solid waste and biosolids. Biotechnology and Bioengineering. 1998; 57(3): 342–355.

26. Karakashev D, Batstone DJ, Angelidaki I. Influence of environmental conditions on methanogenic compositions in anaerobic biogas reactors. Applied and Environmental Microbiology. 2005; 71(1): 331–338.

27. Yu Z, Morrison M, Schanbacher FL. Production and utilization of methane biogas as renewable fuel. In: Vertès, A.A., Blaschek, H.P., Yukawa, H., Qureshi, N. (Eds.), Biomass to Biofuels: Strategies for Global Industries. John Wiley & Sons, Inc., Hoboken, NJ. 2009.

28. Chyi YT, Dague RR. Effects of particulate size in anaerobic acidogenesis using cellulose as a sole carbon source. Water Environ. Res. 1994; 66: 670–678.

29. Han Y, Sung S, Dague RR. Temperature-phased anaerobic digestion of wastewater sludges. Water Sci. Technol. 1997; 36: 367–374.

30. Sakai S, Tsuchida Y, Okino S, Ichihashi O, Kawaguchi H, Watanabe T, Inui M, Yukawa H. Effect of lignocellulose-derived inhibitors on growth of and ethanol production by growth-arrested Corynebacterium glutamicum R. Appl. Environ. Microbiol. 2007; 73: 2349–2353.

31. Bok FA, Plugge CM, Stams AJ. Interspecies electron transfer in methanogenic propionate degrading consortia. Water Res. 2004. 38: 1368–1375.

32. Speece RE, Boonyakitsombut S, Kim M, Azbar N, UrsilloP. Overview of anaerobic treatment: thermophilic and propionate implications. Water Environ. Res. 2006; 78: 460–473.

33. Converti A, del Borghi A, Zilli M, Arni S, del Borghi M. Anaerobic digestion of the vegetable fraction of municipal refuse: mesophilic versus thermophilic conditions. Bioprocess Biosyst. Eng. 1999; 21: 371–376.

34. Van Velsen AFM, Lettinga G, Den Ottelander D. Anaerobic digestion of piggery waste 3: Influence of temperature. Neth. J. Agri. Sci. 1980. 27: 255–267.

35. Iranpour R. Retrospective and perspectives of thermophilic anaerobic digestion: part I. Water Environ. Res. 2006; 78: 99.

36. Chachkhiani M, Dabert P, Abzianidze T, Partskhaladze G, Tsiklauri L, DudauriT, Godon JJ. 16S rDNA characterisation of bacterial and archaeal communities during start-up of anaerobic thermophilic digestion of cattle manure. Bioresour. Technol. 2004; 93: 227–232.

37. Hori T, Haruta S, Ueno Y, Ishii M, Igarashi Y. Dynamic transition of a methanogenic population in response to the concentration of volatile fatty acids in a thermophilic anaerobic digester. Appl. Environ. Microbiol. 2006; 72: 1623–1630.

38. Surata W, Tjokorda G, Tirta NT, Ketut AA. Simple conversion method from gasoline to biogas fueled small engine to powered electric generator. Energy procedia. 2014; 52: 626-632.

39. Samanta A, Das S, Roy PC. Performance analysis of a biogas engine. International J. of Res. in Engg. and Tech. 2016; 5(3): 67-71.

40. Ma AY. Recent advances of anaerobic digestion for energy recovery, Recycling of Solid Waste for Biofuels and Bio-chemicals., Springer, 2016; 87-126.

41. Appels L. Anaerobic digestion in global bio-energy production: potential and research challenges, Renew. Sustain. Energy Rev. 2011; 15(9): 4295-4301.

42. Elasri O, Afilal MEA. Potential for biogas production from the anaerobic digestion of chicken droppings in Morocco, Int. J. Recycl. Org. Waste Agric. 2016; 5(3):195.

43. Jain S. A comprehensive review on operating parameters and different pretreatment methodologies for anaerobic digestion of municipal solid waste, Renew. Sustain. Energy Rev. 2015; 52: 142-154.

44. Ariunbaatar J. Pretreatment methods to enhance anaerobic digestion of organic solid waste, Appl. Energy. 2014; 123: 143-156.

45. Parawira W. Enzyme research and applications in biotechnological intensification of biogas production, Crit. Rev. Biotechnol. 2012; 32(2): 172-186.

46. Christy PM, Gopinath L, Divya D. A review on anaerobic decomposition and enhancement of biogas production through enzymes and microorganisms, Renew. Sustain. Energy Rev. 2014; 34: 167-173.

47. Sreekrishnan T, Kohli S, Rana V. Enhancement of biogas production from solid substrates using different techniques_a review, Bioresour. Technol. 2004; 95(1): 1-10.

48. Wahyudi A, Hendraningsih L, Malik A. Potency of fibrolytic bacteria isolated from Indonesian sheep's colon as inoculum for biogas and methane production, Afr. J. Biotechnol. 2010; 9(20).

49. Sarkar P, Meghvanshi M, Singh R. Microbial consortium: a new approach in effective degradation of organic kitchen wastes, Int. J. Environ. Sci. Dev. 2011; 2(3): 170.

50. Dhadse S, Kankal N, Kumari B, Study of diverse methanogenic and non-methanogenic bacteria used for the enhancement of biogas production, Int. J. Life Sci. Biotechnol. Pharma Res. 2012; 1(2): 176-191.

51. Karthikeyan OP, Heimann K, Muthu SS. Recycling of Solid Waste for Biofuels and Bio-chemicals., Springer, 2016.

52. Astals S, Nolla-Arde'vol V, Mata-Alvarez J. Anaerobic co-digestion of pig manure and crude glycerol at mesophilic conditions: biogas and digestate, Bioresour. Technol. 2012; 110: 63-70.

53. Wang X. Optimizing feeding composition and carbon-nitrogen ratios for improved methane yield during anaerobic co-digestion of dairy, chicken manure and wheat straw, Bioresour. Technol. 2012; 120: 78-83.

54. Mun~oz R. A review on the state-of-the-art of physical/chemical and biological technologies for biogas upgrading, Rev. Environ. Sci. Biotechnol. 2015; 14(4).

55. Sun Q. Selection of appropriate biogas upgrading technology—a review of biogas cleaning, upgrading and utilisation, Renew. Sustain. Energy Rev. 2015; 51: 521-532.

56. Thra¨n D. Biomethane—status and factors affecting market development and trade, IEA Task. 2014: 40.

57. Ryckebosch E, Drouillon M, Vervaeren H. Techniques for transformation of biogas to biomethane, Biomass Bioenergy. 2011, 35(5): 1633-1645.

58. Bauer F. Biogas upgrading—technology overview, comparison and perspectives for the future, Biofuels, Bioprod. Biorefin. 2013; 7(5): 499-511.

59. Andriani D. A review on optimization production and upgrading biogas through CO_2 removal using various techniques, Appl. Biochem. Biotechnol. 2014; 172(4): 1909.

60. Patterson T. An evaluation of the policy and techno-economic factors affecting the potential for biogas upgrading for transport fuel use in the UK, Energy Policy. 2011; 39(3): 1806-1816.

61. Persson M, Jo¨nsson O, Wellinger A. Biogas upgrading to vehicle fuel standards and grid injection, IEA Bioenergy Task, 2006.

62. Basu S. Membrane-based technologies for biogas separations, Chem. Soc. Rev. 2010; 39(2): 750-768.

63. Zornoza B. Mixed matrix membranes for gas separation with special nanoporous fillers, Desalin. Water Treat. 2011; 27(1-3): 42-47.

64. Scholz M, Melin T, Wessling M. Transforming biogas into biomethane using membrane technology, Renew. Sustain. Energy Rev. 2013; 17: 199-212.

65. Makaruk A, Miltner M, Harasek M. Membrane biogas upgrading processes for the production of natural gas substitute, Sep. Purif. Technol. 2010; 74(1): 83-92.
66. Effendi A. Optimising H_2 production from model biogas via combined steam reforming and CO shift reactions, Fuel. 2005; 84(7): 869-874.
67. Kolbitsch P, Pfeifer C, Hofbauer H. Catalytic steam reforming of model biogas, Fuel. 2008; 87(6): 701-706.
68. Lin KH, Chang HF, Chang ACC. Biogas reforming for hydrogen production over mesoporous $Ni_2xCe12xO_2$ catalysts. Int. J. Hydrogen Energy. 2012; 37(20): 15696-15703.
69. Xu G. Producing H_2-rich gas from simulated biogas and applying the gas to a 50W PEFC stack, AIChE J. 2004; 50(10): 2467-2480.
70. Italiano C. Bio-hydrogen production by oxidative steam reforming of biogas over nanocrystalline Ni/CeO_2 catalysts, Int. J. Hydrogen Energy. 2015; 40(35): 11823-11830.
71. Kumar S, Khanal SK, Yadav Y. Proceedings of the First International Conference on Recent Advances in Bioenergy Research., Springer, 2016.
72. Cai X, Dong X, Lin W. Autothermal reforming of methane over Ni catalysts supported on $CuO\_ZrO_2\_CeO_2\_Al_2O_3$, J. Nat. Gas Chem. 2006; 15(2): 122-126.
73. Mosayebi Z. Autothermal reforming of methane over nickel catalysts supported on nanocrystalline $MgAl_2O_4$ with high surface area. Int. J. Hydrogen Energy. 2012; 37(2): 1236-242.
74. Alves HJ. Overview of hydrogen production technologies from biogas and the applications in fuel cells, Int. J. Hydrogen Energy. 2013; 38(13): 5215-5225.
75. San-Jose´-Alonso D. Ni, Co and bimetallic Ni_Co catalysts for the dry reforming of methane, Appl. Catal. A: Gen. 2009; 371(1): 54-59.
76. Bereketidou O, Goula M. Biogas reforming for syngas production over nickel supported on ceria_alumina catalysts, Catal. Today. 2012; 195(1): 93-100.
77. Xu J. Biogas reforming for hydrogen production over nickel and cobalt bimetallic catalysts, Int. J. Hydrogen Energy. 2009; 34(16): 6646-6654.
78. Lau C, Tsolakis A, Wyszynski M. Biogas upgrade to syn-gas (H_2-CO) via dry and oxidative reforming, Int. J. Hydrogen Energy. 2011; 36(1): 397-404.
79. Purwanto H, Akiyama T. Hydrogen production from biogas using hot slag, Int. J. Hydrogen Energy. 2006; 31(4): 491-495.

80. Chun YN. Hydrogen-rich gas production from biogas reforming using plasmatron, Energy Fuels. 2007; 22(1): 123-127.
81. Rueangjitt N, Akarawitoo C, Chavadej S. Production of hydrogen-rich syngas from biogas reforming with partial oxidation using a multi-stage AC gliding arc system, Plasma Chem. Plasma Process. 2012; 32 (3): 583-596.

CHAPTER - 5

PRESENT STATUS AND ADVANCEMENTS IN BIOMASS GASIFICATION

Hitesh Sanchavat[1*], Vinit Modi[2], Tilak V. Chavda[1], Alok Singh[1] and Sandip H. Sengar[1]

[1]*College of Agricultural Engineering and Technology*
Navsari Agricultural University, Dediapada-393040, Gujarat, India
[2]*Center of Excellence for Energy and Environmental Studies*
Deenbandhu Chhotu Ram University of Science and Technology
Murthal, Sonepat-131039, Haryana, India
*Corresponding Author

Biomass gasification is a process of converting solid biomass fuel into a gaseous combustible gas is known as syngas or producer gas through a sequence of thermo-chemical reactions. This chapter explains biomass energy resources availability in India, principle and basic concepts of gasifiers, thermal reactions inside the reactor, fuel characteristics, types of commercially available gasifier, its application, pros and cons design aspects of gasifiers and cost economics. This chapter also mentions la test development in gasifiers in foreign country, development history of gasifiers in India, leading manufacturer of gasifiers in India

1. INTRODUCTION

1.1 Biomass Energy

Biomass is very important natural resource which is produced from photosynthesis reaction and resulted in organic material. Biomass provides a clean, natural energy source that could certainly improve our environment, economy and energy security. Biomass energy generates from biomass compare to far less air emissions than convectional energy sources/ fossil fuels. India being an agrarian country produces significant quantity of biomass in terms of crop residues, crop straw and animal dung. If the huge resources could be managed properly and farmers are empowered on technological interventions, both self-reliance, improves rural economy and prosperity be brought in the rural areas [1].

Photosynthesis process

$$6CO_2 + 6\ H_2O + \text{Sun light} = C_6H_{12}O_6 + 6\ O_2 + 636\ kcal.$$

With serious concern on rising use of fossil fuels, it is important for India to promote renewable /nonconventional energy sources of energy for power fulfilments. India is the 7th largest country in the world spanning 3.29 million hectares [2]. In villages majority of rural household biomass is very important and most utilized renewable energy source.

Biomass energy can be utilized for various applications for thermal application.

1.2 Biomass Energy and Its Sources

It is use of organic material or animal waste used for generate energy (bio fertilizer, electricity or heat). Biomass is obtained from any plant, algae, human or animal derived organic matter are classified as:

- Crop residues, wood from trees and wood factory waste, the construction industry
- Aquatic and Marine biomass: Algae, aquatic weed and water hyacinth etc.
- Animals and animal droppings
- Municipal Solid waste, municipal sewage sludge and the construction industry waste etc.

India is the second largest agro-based economy which generates a large amount of agricultural crop residues. India has an estimated biomass power potential of estimated 17.5 GW in which presently installed capacity as of 31st March 2014 was 1.9 GW including both off-grid and grid connected power plants. For bagasse cogeneration power, India has a potential of estimated 5 GW out of which presently installed capacity as of 31st March 2014 was 2.6 GW. In an agrarian economy like India biomass has huge potential [3]. Energy from biomass is reliable, successful penetration of biomass energy needs effective supply chain in rural as well as small rural based processing industry. There are so many conversion routes of biomass energy into useful thermal application. Development of indigenous conversion technologies in wide range of applications and features utilizing local skill and resources would promote rural employment, empowerment and help in makes self reliable India. A study by International Solid Waste

Association (ISWA) showed that annual worldwide waste generation accounts for 7-10 billion tones in total, of which approximately 2 billion tones are made up of municipal solid waste (MSW). The typical waste are made up of 24 percent MSW, 36 percent construction and demolition (C&D) waste, 21 percent industrial waste, 11 percent commercial waste and 5 percent arises from wastewater treatment. Only three percent of the total solid waste is presently used for as useful energy and there is a significant huge potential to use the left over waste for energy recovery [4].

1.3 Waste Types and Composition

Waste is a general term and encompasses a large variety of materials originating from various sources associated with human development. Waste is heterogeneous in nature [5] and very tedious task to separate down to individual waste types. Easily divisible waste streams are sorted and separated at a material improvement facility (MRF), and remaining residue waste that is difficult and too expensive to separate is leave for energy recovery. Table 5.1 provides ravage types and their various sources variety of waste materials in the MSW. Country to country composition of MSW varies and can having high quantity of moisture content up to 50 w/w %.

Table 5.1: Waste, its Source, Types and Composition [6].

Source	Type	Composition
Municipal solid waste (MSW)	Residential	Food wastes, paper, cardboard, polythene, textiles, leather, wood, household hazardous wastes, glass, metals, ashes, special wastes (e.g., bulky items, consumer electronics, batteries, oil, tyres), e-waste.
	Industrial	Housekeeping wastes, packaging, food wastes, oil, wood, steel, concrete, bricks, ashes, hazardous wastes.
	Commercial & institutional	Paper, cardboard, polythene, wood, food wastes, glass, metals, special wastes, hazardous wastes, e-waste.
	Construction & demolition	Wood, steel, concrete, soil, bricks, tiles, glass, plastics, insulation, hazardous waste.
	Municipal services	Street sweepings, landscape & tree flourishes, mud, wastes from fun areas.
Process waste		Scrap material, slag, topsoil, process fluid & chemicals, waste rocket
Medical waste		Infectious wastes (bandage, gloves, cultures, bottles, blood & bodily fluids, skin), hazardous wastes (sharps instruments like needles, chemical), radioactive and pharmaceutical wastes.

[Table Contd.

Contd. Table]

Source	Type
Agricultural waste	Spoiled food wastes, pulse industry waste, rice husks, sugarcane trash, yarn (cotton) stalks, coconut shells, pesticides, animal dung, soiled water, silage waste matter, plastic, scrap machinery, veterinary medicine.

As shown in above Table 5.1, the diversity of materials in the waste and complexity of separation leads to a variety of waste treatment options. Since no single waste treatment option can suitable for all types of waste, hence different waste types to need different types of treatments. Source segregation is most effective method, it possible to separate waste streams that can be destined for a waste treatment facility.

The present availability of biomass in India is approximate 500 million metric tons per year. Studies sponsored by the MNRE has approximate surplus biomass available is 120-150 million metric tons per annum covering agricultural and forestry residues equivalent probable 18,000 MW [7]. Table 5.2 show the state-wise crop residue generated, residue surplus and burned in India.

Table 5.2: State-Wise Crop Residue Generated, Residue Surplus and Burned [8]

(*Crop residue in Million Tonne)

S. N.	States	Residue Generation*	Residue surplus*	Residue burned*
1.	Andhra Pradesh	43.89	6.96	2.73
2.	Arunachal Pradesh	0.40	0.07	0.04
3.	Assam	11.43	2.34	0.73
4.	Bihar	25.29	5.08	3.19
5.	Chhattisgarh	11.25	2.12	0.83
6.	Goa	0.57	0.14	0.04
7.	Gujarat	28.73	8.90	3.81
8.	Haryana	27.83	11.22	9.08
9.	Himachal Pradesh	2.85	1.03	0.41
10.	Jammu &Kashmir	1.59	0.28	0.89
11.	Jharkhand	3.61	0.89	1.10
12.	Karnataka	33.94	8.98	5.66
13.	Kerala	9.74	5.07	0.22
14.	Madhya Pradesh	33.18	10.22	1.91
15.	Maharashtra	46.45	14.67	7.42

[Table Contd.

Contd. Table]

S. N.	States	Residue Generation*	Residue surplus*	Residue burned*
16.	Manipur	0.90	0.11	0.07
17.	Meghalaya	0.51	0.09	0.05
18.	Mizoram	0.06	0.01	0.01
19.	Nagaland	0.49	0.09	0.08
20.	Orissa	20.07	3.68	1.34
21.	Punjab	50.75	24.83	19.65
22.	Rajasthan	29.32	8.52	1.78
23.	Sikkim	0.15	0.02	0.01
24.	Tamil Nadu	19.93	7.05	4.08
25.	Tripura	0.04	0.02	0.02
26.	Uttarakhand	2.86	0.63	0.78
27.	Uttar Pradesh	59.97	13.53	21.92
28.	West Bengal	35.93	4.29	4.96
	Total	**501.73**	**140.84**	**92.81**

1.4 BIOMASS PROPERTIES

Thermal processes processing of biomass depends on principle constitute from it form. Cellulose is a linear polymer of hydro glucose units, hemi-cellulose is a combination of polymers of 5- and 6-carbon a hydro sugar and lignin is an irregular polymer of phenyl propane. In biomass, these three polymers form an interpenetrating system or block co polymer that varies in composition across the cell wall. Nevertheless, in huge samples, there is a relatively stable atomic ratio of $CH_{1.4} O_{0.6}$. (The ratios will vary slightly with species. Coal is normally about $CH_{0.9} O_{0.1}$ but vary more commonly in composition.) The link between solid, liquid, and gaseous fuel is shown in Fig. 5.1 (a) where the comparative atomic concentrations of carbon, hydrogen, and oxygen are plotted for a range of fuels.

Here it is observe that the solid fuels, biomass, coal and charcoal, lie in the lower left part of the diagram; liquid and gaseous hydrocarbon fuels lie in the upper left part; CO and H_2 are connected by the bisector of the triangle and the ignition products of fuels, CO_2 and H_2O, lie on a vertical line on the right part. The input to gasification is variety of solid carbonaceous material, i.e. biomass or coal. All organic carbonaceous material consists of carbon (C), hydrogen (H), and oxygen (O) atoms. The goal in gasification is to split down the organic material into the plain fuel gases of hydrogen (H_2) and carbon monoxide (CO). Carbon monoxide and hydrogen have about the same energy density by volume. Both provides very clean burning as they only need one oxygen atom in one simple step to arrive at the proper end states of combustion, where the reaction releases CO_2 and H_2O. This is why an engine run on producer gas offers clean emissions [10].

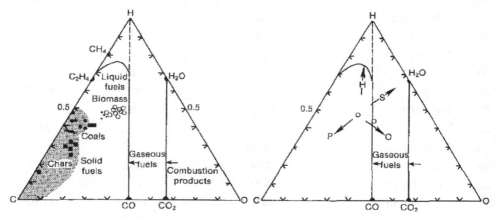

Figure 5.1: (a) Phase Diagram Showing the Relative Proportions of Carbon, Hydrogen, and Oxygen in Solid, Liquid, and Gaseous Fuels **(b)** Chemical Changes during Biomass Conversion Processes [9]

2. BIOMASS ENERGY CONVERSION TECHNOLOGIES

The term "conversion technology" encompasses a wide range of technologies used to convert solid waste into clean fuel of high calorific value and organic fertilizer. Wide range of bio conversion technologies as shown in Fig. 5.2 suitable for rural applications are as under thermal conversion processes for biomass involve some of these or all of the following processes [9].

a. **Combustion:** Biomass + Stoichiometric Oxygen = Hot combustion products
b. **Pyrolysis:** Biomass + Heat = Charcoal, Oil, Gas
c. **Gasification:** Biomass + Limited amount of Oxygen = Fuel gas

2.1 Biomass Gasification

Biomass gasification is a complex process involving drying the feedstock follows by Pyrolysis, fractional combustion of intermediates and to finish gasification of the resulting products. It is perform in the occurrence of a gasifying media which can be air, oxygen (O_2), steam (H_2O) or carbon dioxide (CO_2), inside a reactor called a gasifier. The calorific value of the product gas is affected by on the gasifying agent. The product gas from air gasification gives a heat value of approximately 4-7 MJ/m^3 whereas when gasifying utilizing pure O_2, the heat value of approximately 12-28 MJ/m^3 [11] . The quality and properties of the produce are dependent on the feedstock material, gasifying agent, feedstock size, temperature and pressure inside the gasifier reactor, design of reactor, mechanism and sorbent [12]. Syngas or producer gas is used for rural electrification, dairy industry for process heat or promoting village level food processing industries. Farmers can generate income from sell of surplus biomass material. The benefit of gasification is that using the syngas is potentially more efficient compare to direct combustion of the original fuel.

Almost any type of organic substance can be used as the raw substance for gasification, such as wood, biomass and plastic waste. Gasification is like a choked (clogged) combustion of solid organic material. It is truly understood as staggered combustion. It is a series of different thermal events put together. Though gasifiers are relatively simple devices, successful operation of gasifiers and cleaning of gas however is not so simple. No neat system exists because the thermodynamics of gasifier process are not well understood. Globally and even in India, there are many researchers have been done and are undergoing. Mature commercial gasifiers are being made available by many units for vivid application of process heat and electrical power generation.

2.2 Biomass Gasification Process

Gasification is completed by separate thermal processes: Drying, Pyrolysis, Combustion and Reduction as shown in Fig. 5.3 and 5.4.

All of these processes are completed in presence in the flame; though they mix in a manner the process is invisible to eyes. Gasification is the technology to separate out the processes, so that the resulting gas is routed out through distinct outlet. The figure below illustrates the main five process occur in the gasifier.

2.2.1 Drying

Drying is removal of moisture from the biomass. The moisture needs to be removed from the fuel before any above 100°C processes happen. To lead successful gasification, moisture in the biomass must get vaporized at some point in the higher temp processes. High moisture content feedstock or poor handling of the moisture internally, is one of the most general reasons for failure to produce clean gas. It has been observed from research that approximate 9% of energy lost for reducing moisture content from 30 to 9 percent and reduction in calorific value around 26% [1].

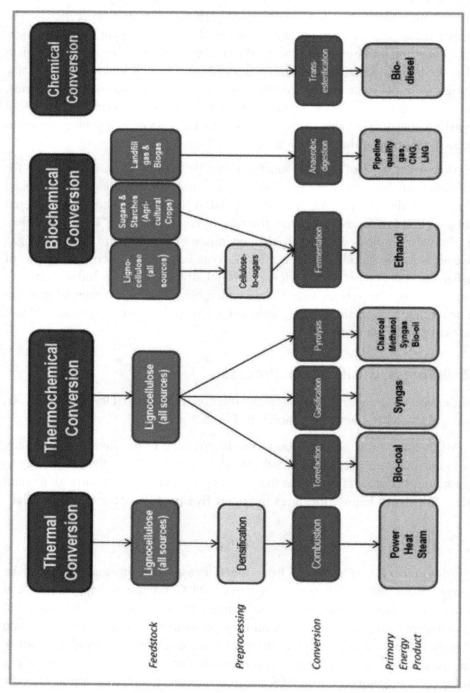

Figure 5.2: Flow Chart of Biomass Energy Conversion Technologies

2.2.2 Biomass Pyrolysis

Pyrolysis is the breaking down (pyrolysis) of a substance by heat (pyro). Pyrolysis is the application of heat to feed stock, in an absence of air (O_2), so as to fracture it down into charcoal and various tar gas and liquids. It is essentially the route of charring. Feed stock begins to rapidly decompose with heat once its temperature reaches roughly 240°C. The feed stock breaks down into a combination of solids, liquids and gasses. The solids that remain called as charcoal and the gasses and liquids that released are together known as tars.

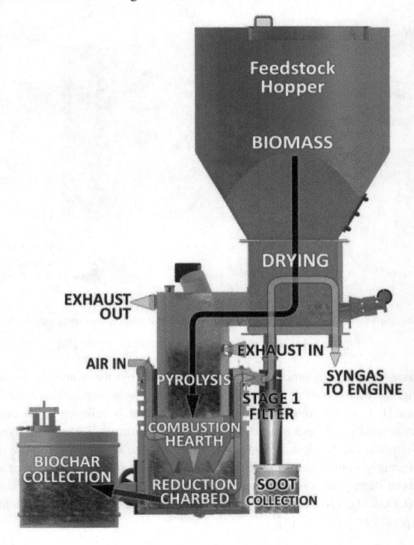

Figure 5.3: Schematic of Gasification Process [13]

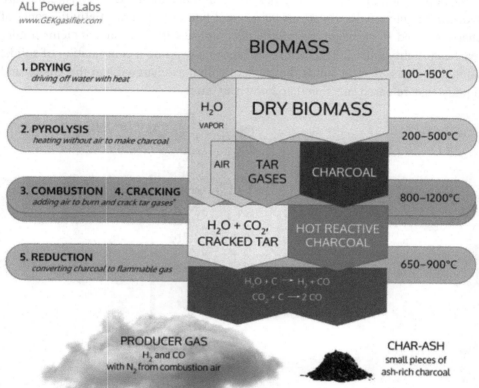

Figure 5.4: Five Processes inside Gasifier Reactor [13]

The gasses and liquids produced during lesser temp pyrolysis are just portion of the original biomass that breaks off with heat. These portions are the more complex H, C and O molecules in the biomass that is referred to as volatiles (substance easily gets evaporated at normal temperatures). During the pyrolysis process, when feedstock is heated in the nonexistence of oxygen, volatile matter and inherently bound water in the fuel are evaporated off to form a vapor which consists of largely tar, oil, and gases. The pyrolysis process derives charcoal (C), gases (CO, CO_2, H_2, H_2O, CH_4), and tar vapors (with an approximate atomic makeup of $CH_{-1.2}O_{0.5}$) [9].

2.2.3 Combustion

This is the most easily understood of the five Processes of gasification. In the gasification process, the feed stock and the gasification media (air, oxygen, steam or a mixture of these) are pour into the gasifier. The oxygen or air is part of the feed stock fuel in order to heat the incoming fuel to the process temperature and to give the result heat required affected by the charge of flow of feed stock (biomass). Combustion is the only net exothermic process as all of the energy or heat that is required for drying, pyrolysis, and reduction comes also directly from combustion or is recovered indirectly through gasifier by heat exchange processes from combustion.

2.2.4 Cracking

Cracking is the process of breaking down big complex molecules such as tar into lighter gases through exposure to heat. This process is critical for the production of clean gas that is well-suited with an internal combustion engine because tar gases condense into sticky tar that will quickly foul the valves of an engine. Cracking is also essential to ensure proper combustion because whole combustion only occurs when combustible gases properly mix with oxygen. In the route of combustion, the high temperatures produced decompose the large tar molecules that pass during the combustion zone.

2.2.5 Reduction

Reduction is the process of separating oxygen atoms, hydrocarbon (HC) molecules of combustion process, so as return molecules to forms that can burn again as shown in Fig. 5.5. Reduction is the direct reverse process of combustion. Combustion is the combination of combustible gases with oxygen to release heat, producing water vapor and carbon dioxide as waste products. Reduction is the removal of oxygen from these waste products at high temperature to produce combustible gases. Combustion and reduction are the same and reverse reactions. In fact, in the majority burning environments, they are both operating simultaneously, in some form of dynamic equilibrium, with frequent movement back and forth between the two processes.

Reduction in a gasifier is accomplished by passing carbon dioxide (CO_2) or water vapor (H_2O) crossways a bed of red hot charcoal (C). The carbon in the hot charcoal is highly reactive with oxygen, it has such a high oxygen resemblance that it strips the oxygen off water vapor and carbon dioxide and redistributes it to as many single bond sites as possible. The oxygen is additional attracted to the bond site on the C than to itself, thus no free oxygen can survive in its usual diatomic O_2 form. All available oxygen will bond to available C sites as individual O until all the oxygen is gone. When all the existing oxygen is redistributed as single atoms, reduction stops. Through this process, CO_2 is reduced by carbon to produce two CO molecules, and H_2O is reduced by carbon to produce H_2 and CO. Both H_2 and CO are combustible fuel gases and those fuel gasses can then be transferred do desired work elsewhere.

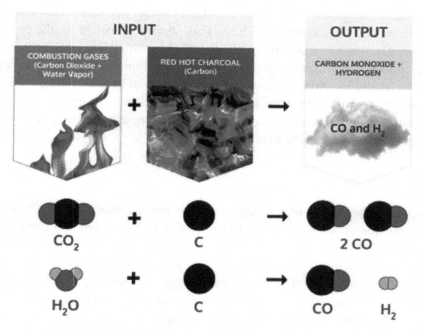

Figure 5.5: Reactions in the Reduction Zone [13]

2.3 Chemistry of Biomass Gasification

Gasification takes place at elevated temperature in the presence of an oxidizing agent (also called a gasifying agent). Heat is supplied to the gasifier either directly or indirectly which raises the gasification temperature of 600–1500°C. Oxidizing agents are normally air, steam, nitrogen, carbon dioxide, oxygen or a combination of these. In the presence of an oxidizing agent at elevated temperature, the big polymeric molecules of biomass decompose into lighter molecules and finally to permanent gases (CO, H_2, CH_4 and lighter hydrocarbons), ash, char, tar and minor contaminants. The following reaction occurs in gasification process [14]:

CH_xO_y (biomass) + O_2 (21% of air) + H_2O (steam)
= CH_4 + CO + CO_2 + H_2 + H_2O (unreacted steam) + C (char) + tar ... (1)
2C + O_2 = 2CO (partial oxidation reaction) ... (2)
C + O_2 = CO_2 (complete oxidation reaction/combustion) ... (3)
C + $2H_2$ = CH_4+75000kJ/k-mol (methane reaction) ... (4)
CO + H_2O = CO_2 + H_2 (water shift reaction) ... (5)
CH_4 + H_2O = CO + $3H_2$ (steam reforming reaction) ... (6)
C + H_2O = CO + H_2 (water gas reaction) ... (7)
C + CO_2 = 2CO -172600kJ/kg mol (Boudourd reaction) ... (8)

The overall reaction in an air and steam gasifier can be represented by Equation 1, which proceeds with multiple reactions and pathways. Equations 2–8 are common reactions involved during gasification. Among these, Equations 4–7 happen when steam is available during gasification [14].

2.4 Composition of Producer Gas

Biomass gasification process produces Syngas or Producer gas. Syngas is a mixture of Carbon Monoxide and Hydrogen which is the product of steam or oxygen gasification of organic material such as biomass. If the gasification product contains significant amount of non-combustible gases such as nitrogen and carbon dioxide, then the resultant gas is known as producer gas. Tables 5.3 and 5.4 below shows the typical composition of syngas and producer gas derived from biomass gasification process, Table 5.5 presents typical composition of producer gas for various kinds of feedstock [1 & 9].

Table 5.3: Typical Composition of Syngas from Biomass Gasification

Compound	Gas (Vol. %)
Carbon monoxide (CO)	46.0
Carbon dioxide (CO_2)	25.0
Hydrogen (H_2)	29.0
H_2/CO	0.636
H_2/CO_2	1.167

Table 5.4: Typical Composition of Producer Gas from Biomass Gasification [1 & 9]

Compound	Gas (Vol. %)	Dry gas (Vol. %)
Carbon monoxide (CO)	21.0	22.1
Carbon dioxide (CO_2)	9.7	10.2
Hydrogen (H_2)	14.5	15.2
Water (H_2O)	4.8	-
Methane (CH_4)	1.6	1.7
Nitrogen (N_2)	48.4	50.8

Gas High Heating Value: 4.5 to 6MJ/m^3
Air Ratio Required for Gasification: 2.38 kg wood/kg air (lb/lb)
Air Ratio Required for Gas Combustion: 1.15 kg wood/kg air (lb/lb)
(These values are based on ash-and moisture-free bio-mass)

3. CLASSIFICATION OF GASIFIER

In a gasifier, fuel interacts with air or oxygen and steam. So the gasifiers are classified as per the way air or oxygen is introduced in it. The choice of the one type of gasifier compare to the other is mostly determined by the feedstock, its final available form, its size, moisture content and ash content. On a bigger scale there are following types of gasifiers largely available for commercial applications [16].

a. Updraft Gasifier
b. Downdraft Gasifier
c. Cross Draft Gasifier
d. Fluidized Bed Gasifier
e. Twin Fire Gasifier
f. Other Gasifier

3.1 Updraft Gasifier

The updraft gasifier (Fig. 5.6) is widely used for coal gasification and nonvolatile fuels such as charcoal. However, the high rate of tar production (5-20%) makes them impractical for high volatile fuels where a clean gas is required. [1,14].

3.2 Downdraft Gasifier

The downdraft gasifier (Fig. 5.6,) was developed to convert high volatile fuels (wood, biomass) to low tar gas and therefore has proven to be the most successful design for power generation.

Table 5.5: Typical Composition of Producer Gas in Various Feedstocks

Compound	Wood (Vol., %)	Corn cobs (Vol., %)	Barley straw (Vol., %)	Tree pruning (Vol., %)	Rice straw (Vol., %)	Peat (Vol., %)
CO	23.90	21.70	20.90	18.80	26.10	16.15
CO_2	09.70	10.20	10.90	13.70	08.40	15.30
H_2	16.30	16.90	13.40	16.40	12.40	12.30
CH_4	08.17	04.46	04.94	04.75	07.79	00.75
C_2H_6	00.43	00.23	00.26	00.25	00.41	-
N_2	41.50	46.51	49.60	46.10	44.90	55.50

(These values are based on ash-and moisture-free bio-mass) [15]

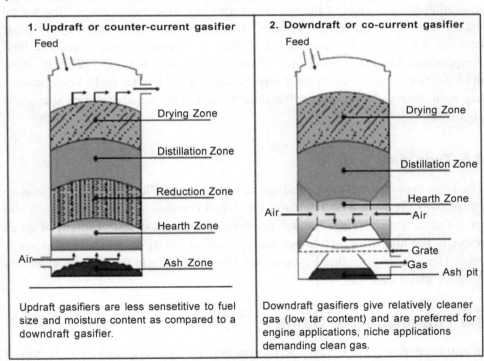

Figure 5.6: Diagram of Updraft and Downdraft Gasification [17]

3.3 Fluidized Bed Gasifier

Fluidized beds are favored by lots of designers for gasifiers producing more than 40 GJ (th)/h* [40 MBtu (th)/h] and for gasifier using minor particle feedstock sizes. In a fluidized bed as shown below in Fig. 5.7, air raises through a grate at elevated enough velocity to climb the particles above the grate, therefore forming a "fluidized bed". Above the bed itself the vessel increases in diameter, decreasing the gas velocity and causing particles to recirculate inside the bed itself. The recirculation results in elevated heat and mass transfer between particle and gas stream.

3.4 Cross Draft Gasifier

The cross draft gasifier shown in Fig. 5.8 is the simple and light gasifier. Air enters at elevated velocity through a single nozzle, induces substantial circulation, and flows across the bed of fuel and char. This produces very elevated temperatures in a very small volume and results in production of a low-tar gas, permitting rapid adjustment to engine load changes. The fuel and ash provide as insulation for the walls of the gasifier, permitting mild-steel construction for all parts excluding the nozzles and grates, which may need refractory alloys or some cooling. Air-cooled or water-cooled nozzles are frequently requisite. The high temperatures reached need a low-ash fuel to prevent slogging [18]. The cross draft gasifier is usually considered appropriate only for low-tar fuels. Some success has been observed with un-pyrolyzed biomass, but the nozzle-to-grate spacing is significant [9].

3.5 Twin Fire Gasifier

The advantages of co-current and counter- current gasifier are combined in a so called twin fire gasifier as shown in Fig 5.9. It consists of two zones drying and gasification zone. Drying, low temperature carbonization and cracking of gases occur in the high zone, while permanent gasification of charcoal takes in lower zone. Here gas temperature lies between 460-520 °C. Twin fire gasifier generally produce clean gas.

3.6 Entrained-Flow Gasifier

In entrained-flow gasifier (Fig. 5.10), fuel and air are introduced from the top of the reactor, and fuel is carried by the air in the reactor. The operating temperatures are 1200–1600 °C and the pressure is 20–80 bar. Entrained-flow gasifier can be used for any type of fuel so long as it is dry (low moisture) and has low ash content. Due to the short residence time (0.5–4.0 seconds), high temperatures are required for such gasifiers. The advantage of entrained-flow gasifiers is that the gas contains very little tar. However, thermal efficiency is rather lower as the gas must be cooled before it can be cleaned.

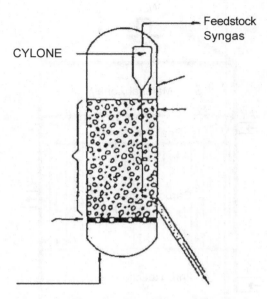

Figure 5.7: Fluidized Bed Gasifier

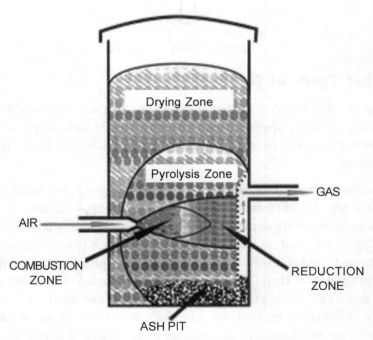

Figure 5.8: Diagram of Cross Draft Gasifier [19]

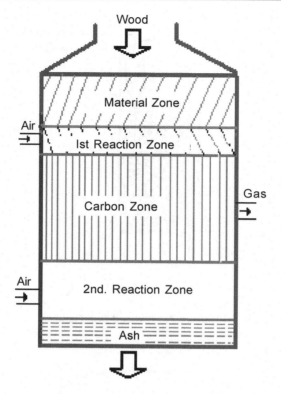

Figure 5.9: Twin Fire gasifier

3.7 Other Types of Gasifiers

A number of other biomass gasifier systems (double fired, entrained bed, molten bath), which are partly spin-offs from coal gasification technology, are currently under development and latest advanced gasifiers which used in foreign country are as follows:

Currently, in biomass gasification plants, clean gas is produced at ambient temperature after filtration and scrubbing, limiting its applications. The reduction in gas temperature owing to cleaning followed by conditioning reduces the overall profitability of the plant. Moreover, if the tar separation is not very effective, the gas quality and yield will suffer, making it unfit for applications where high levels of purity are essential. Therefore, gas conditioning preceded by clean-up at elevated temperatures (i.e., "hot gas cleanup", HGCU) is necessary, to ensure high efficiency in industrial applications, especially in the case of steam gasification. Unique gasification technology investigated by research and development (R&D) establishments and industries in Europe and the US has made it possible to have immediate and efficient conversion of the outlet gas. They are used in fuel cells and micro gas turbines along with power plants [20]. An example of a novel HGCU process is the utilize of plasma torches to crack tars; this differs from plasma gasification where the plasma is used for energy generation by gasifying biomass, MSW as well as refuse derived fuel (RDF) [22,23].

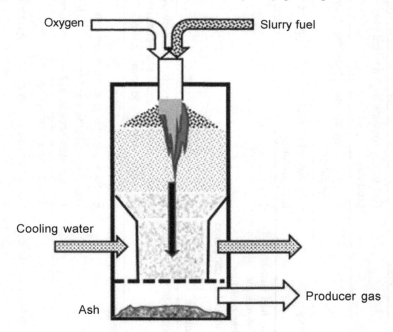

Figure 5.10: Entrained Flow Gasifier

3.8 Plasma Gasification for Toxic Organic Waste

Plasma is one of the fundamental states of matter and can be generated either by heating a gas or by exposing it to a strong electromagnetic (EM) field. There are two types of plasma-thermal plasma and cold plasma. Thermal plasma is created at ambient pressure while cold plasma is produced in a vacuum. Thermal plasma is generally produced with gases such as argon (Ar), N_2, H_2, H_2O vapor or a gas mixture at a temperature of around 4700 °C or higher. In plasma gasification, AC or DC arc plasma torch generators are used [23]. Plasma is utilized in two different way in the gasification process: (1) Plasma is used as a heat provider during gasification; (2) plasma is used for tar cracking after standard gasification. Primarily, plasma gasification is employed for the decomposition of toxic organic wastes, along with rubber and plastics, although the first reason and currently the main application for plasma gasification is the treatment of hazardous biomass waste. However, the technology has also gained interest for syngas production and electricity generation

Table 5.6: Advantages and Disadvantages of Different Types of Gasifier [21]

Gasifier type	Advantages	Disadvantages
Updraft	Simple design	Gas output contains high amount of tar and pyrolysis products
	High charcoal burn-out	Extensive gas cleaning required if used for power application
	High fuel to gas conversion efficiency	Internal heat exchange leading to low gas exit temperatures
	Fuel flexibility in terms of feedstock, fuel switching does not require any changes in the reactor. Tolerates higher ash content, higher moisture content and greater size variation in fuel as compared to downdraft gasifiers	
	More suitable with thermal applications	
Downdraft	Gas output is with less tar, so more suitable for use in gas engines	Producer gas gets contaminated easily by ash and other fine particles, and a separation device, requires two-stage cyclone and ceramic filter.
	Suitable for small to medium scale power application (100 kWth–5 MWth).	Does not permit fuel flexibilities
		Produce more tar at low temperature
		The gas leaves the gasifier at a high temperature, so dedicated cooling system is required.
Cross draft	Economically viable for small-scale application	High amount of tar produced
	Due to high temperatures, gas cleaning requirements are low start-up time is relatively short	Doesn't handle fuel that has a high tar content

[Table Contd.

[Contd. Table]

Gasifier type	Advantages	Disadvantages
Fluidized bed gasifier	Compact construction	Gas stream contains fine particles of dust
	Uniform temperature profile	Complex system due to low biomass hold up in the fuel bed
	Feedstock flexibility and ability to deal with fluffy and fine grained materials without the need of pre-processing.	Variety of biomass can be used but accepts biomass of 0.1 cm to 1 cm size
	High ash melting point of biomass does not lead to clinker formation	
Entrained-flow	Applicable to large systems	High investment
	Short residence time for biomass	Strict fuel requirements

in recent years as the costs have entered into a commercially competitive range. A plasma gasification plant (Fig. 5.11) at Utashinai, Japan has been in service since 2002 and as of 2014, gasifies the large quantity of municipal solid waste per day [25] and produces 7.9 MWh electricity [24].

Figure 5.11 shows a plasma gasifier where the reactor chamber is linked to a non-transferred DC arc plasma torch generator [26]. Due to the very elevated temperatures produced it can be employed for toxic wastes, rubber and plastic treatment. Energy is simultaneously produced from biomass gasification (BG). Though this concept was originally designed for municipal and other waste treatment, it was later extended for high-quality syngas generation. At elevated temperature, gasification of feedstock occurs in milliseconds [27]. The key purported benefits of this process are syngas yield with high H_2 and CO content, improved heat content, low CO_2 yield and low tar content.105,106 The process is working for wet biomasses such as sewage sludge which are otherwise difficult to gasify, and small effect of particle dimension and structure of feedstock is noted. Major limitations are high construction and maintenance costs because of the high electricity consumption to generate plasma, resulting in low overall efficiency. For instance, a base case situation with a 680 tonne per day waste gasification plant

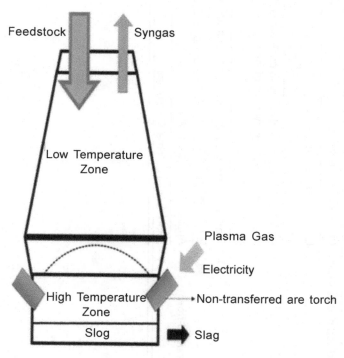

Figure 5.11: Schematic of Plasma Gasifier [26]

which would be suitable for a small town or regional facility, would cost an estimated Rs.7700 million to construct, which is almost three times compare to cost of other waste treatment facilities (e.g. incineration).

3.9 Supercritical Water Gasification (SCWG) for Wet Biomass

Water above its critical point (T = 374.12 °C and P = 221.2 bar) is identified as supercritical. Under these settings, the liquid and gas phases do not be present and supercritical water shows typical reactivity and solvency characteristics. Solubility's of organic materials and gases which are generally insoluble are enhanced. The properties of supercritical water lie between those of the liquid and gaseous phases. SCWG has been applied to wet biomass without the need for pre-drying, which is a major advantage over other more conventional gasification techniques. Many investigations on various feed stocks such as agricultural wastes, leather wastes, switch grass, sewage sludge, algae, manure and black liquor have been conducted [28 -34] . Employing SCWG, liquid biomass such as olive mill water can be used with the production of low-tar H_2 gas [29]. A simplified schematic of a SCWG setup is shown in Fig. 5.12.

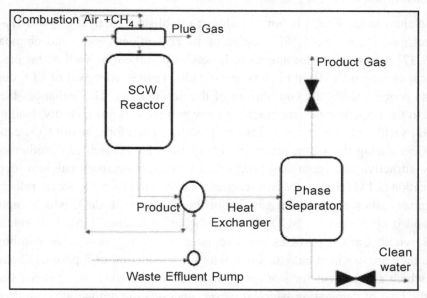

Figure 5.12: Simplified Schematic of SCWG [32]

Product gas from SCWG mostly comprises H_2, CO_2, CH_4 and CO. The CO yield is relatively low as CO transforms into CO_2 during the water–gas-shift reaction. [34] Tar and coke formation are shortened by rapid dissolution of product gas components in supercritical H_2O. Guo et al. [35] and Feng et al [36] found that over 600 °C, H_2 is the dominant component of the produced gas, since H_2O is a tough oxidant which reacts with carbon to release H_2 and CO, whereas CH_4 is the main element below 450 °C. Heating of H_2O to the reaction temperature necessitates a great amount of energy input. However, employing appropriate catalysts can lower the reaction temperature. This reduces the operational and equipment cost and increases conversion efficiency and H_2 production.

4. SORPTION-ENHANCED REFORMING (SER) AND BIOMASS GASIFICATION WITH CO_2 CAPTURE

In steam reforming of biomass, separation of H_2 from a product gas containing CO_2 and tar incurs large cost penalties. Therefore, a solution where the CO_2 and tar produced during steam reforming are simultaneously captured has significant potential to make the process more cost-effective [37,38]. The primary method employs catalysts inside the gasifier while the secondary method uses them downstream. [39,40] Although the secondary method is additional effective, the primary method has gained extra attention on account of avoiding intricate downstream clean-up [41].

Calcium oxide (CaO) is now an almost established catalyst to yield H_2-rich product gas; [42,43,44,45,46] because of its cost effectiveness and abundance [37,39,47] it has gain a lot attention. It acts as a sorbent as well as tar cracker and heat carrier in fluidized biomass gasification (FBG). Removal of CO_2 during the BG process shifts the equilibrium of the product gas. This enhances the H_2 yield. In the same manner, tar cracking increases the exit gas quantity, leading to high H_2 yield and conversion efficiency [48,49] . Therefore, in situ CO_2 capture with CaO during the steam reforming of biomass for H_2-rich gas production is highly attractive and promising [50,51,52,53,54]. CaO-assisted calcium looping gasification (CLG) consists of two reactors as shown in Fig. 5.13. Steam reforming of biomass takes place in the gasifier in the presence of CaO, which captures CO_2 and is converted to $CaCO_3$ via the carbonation reaction [55]. This enhances the H_2 yield. $CaCO_3$ particles are circulated to the regenerator or combustor, where they are calcined back to CaO, with the production of a pure CO2 stream [56], which can be sent for storage. CaO is recycled back to the gasifier beside with the heat of calcination which it carries and aids in compensate endothermic reactions in the gasifier [50] .Therefore, this is a low-energy requirement and eco-friendly process of H_2 production with enhanced efficiency of H_2 production.

Since CaO captures CO_2 according to the carbonation reaction [55], it will lead to a reduction in the partial pressure of CO_2 under gasification conditions.

This reduction in CO_2 partial pressure drives the water-gas-shift reaction [57] forward in accordance with Le Châtelier's principle. This leads to an increased yield of H_2 [51]. Later CaO is recovered by calcination [33]. The efficacy of the reaction is a subset of other parameters also, such as steam-to-biomass ratio (S/B), temperature, pressure, and the amount of CaO. Advantage of the process is high H_2 volume with low tar content even at lower temperatures in the presence of CaO, thus making it a desirable choice for a sorbent in the steam reforming of biomass. A major drawback of using a CaO sorbent in steam-assisted BG is unbalanced H_2 production due to deactivation of CaO through the regeneration. Although CaO is potentially capable in tar reforming and CO_2 capture, the process would not be reasonably viable if the CaO could not be regenerate after the carbonation reaction. Consequently, the supply of CaO must be replenished [58]. In order to overcome this problem to some extent, calcium looping gasification (CLG) was introduced.

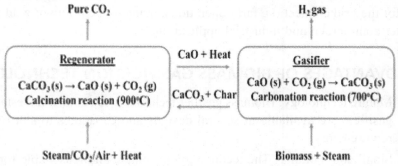

Figure 5.13: CaO Looping in SER

5. APPLICATION OF SYNGAS/PRODUCER GAS

The gasifier gas can be used either to produce heat or to generate electricity.

5.1 Thermal Applications

Producer gas can also be burnt directly in open air, much like Liquid Petroleum Gas (LPG), and therefore can be used for cooking, boiling water, producing steam, and drying food and other materials.

Key end applications of the gas are:

a. **Dryer:** The hot gas after combustion can be mixed with the right quantity of secondary air to lower its temperature to the desired level for use in dryers in the agro industries such as drying of fruits, vegetables, tea, spices, medicinal and herbal produce. It is also suitable for dairy industries for production of milk powder, casein etc, seasoning of wood and timber, drying of textile materials.

b. **Kilns:** Firing of tiles, pottery articles, limestone and refractories, where temperatures of 800–950°C are required.

c. **Boilers**: Producer gas can be used as fuel in boilers to produce steam or hot water.

5.2 Power Applications

Producer gas have been utilizes for generating motive power to run either dual-fuel engines which run on a mixture of gas and diesel by substitution of up to 85% of diesel or engines that run on producer gas alone (100% diesel replacement). Normally the fuel-to-electricity efficiency of gasification is 35%–45%, which is more than that of direct combustion (10%–20%). Generated electricity can be utilize for the grid connection, farm operations, irrigation, chilling or cold storage and other commercial and industrial applications.

6. ADVANTAGES OF BIOMASS GASIFICATION TECHNOLOGIES

a. Mature technology: Biomass gasifier technology is commercial established gasifiers are available in several designs and capacities to suit different requirements.

b. Small and modular: The technology is suitable and reasonable for small, decentralized applications, typically with capacities lesser than a megawatt.

c. Flexible operation: A gasifier based power system, can generate electricity when necessary and also wherever required.

d. Biomass gasifier based systems can be set up at almost several places wherever biomass feedstock is available.

e. Economically viable: For small-scale systems, the cost of power generation by biomass gasification technology is lower than that of conventional diesel based power generation.

f. Socio-economically beneficial: Biomass gasifier based systems creates employment for large village workers.

g. Mitigate climate change: Biomass is a CO_2 neutral fuel and, therefore, unlike fossil fuels such as diesel does not contribute to net CO_2 emissions, which makes biomass based power generation systems an attractive option in mitigating the adverse effects of climate change.

7. COOLING AND CLEANING SYSTEM OF GASIFIER

Trouble free function of an internal combustion engine using producer gas as fuel requires a comparatively clean gas. Gas cooling primarily serves the purpose of increasing the density of the gas in order to maximize the quantity of combustible gas entering the cylinder of the engine at each stroke. Cooling system as shown in Fig. 5.14 uses a shell and tube heat exchanger to cool the gas stream from the gasification plant.

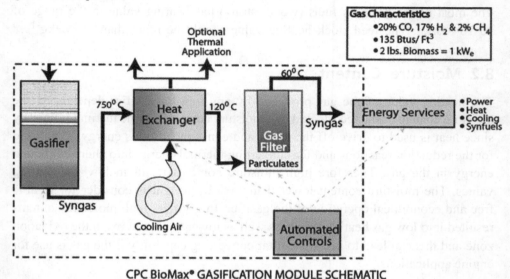

Figure 5.14: Cooling and Cleaning System of Gasifier

A ten percent temperature decrease of the gas increases the maximum output of the engine by approximately two percent. Cooling also contributes to gas cleaning and makes it promising to avoid condensation of moisture in the gas following it is mixed with air before the engine intake. Waste heat from the gas cooling is used to dry feedstock. Liquid cooling or scrubbing of the gas is also practiced for cooling of the gas, which needs flow of water. The char/ash particles are detached by self-cleaning, passive bag-type filters and mechanically stored in collection bags. Depending on the feedstock, the improved char/ash can be used as a soil alteration.

8. FACTORS AFFECTING PERFORMANCE OF THE GASIFIER

Input fuel characteristic play important role in the performance of any gasifier, key factors which govern the performance of gasifier are [15]:

8.1 Energy Content of the Feedstock

The preference of a fuel for gasification is decided by its heating value. Energy content influences the efficiency of a given gasification system. The simply reasonable way of presenting fuel heating values for gasification purposes is to lower heating values (excluding the heat of condensation of the water produced) on an ash inclusive basis and with specific reference to the actual moisture content of the fuel. Fuel with higher energy content is always better for gasification. The most of the biomass fuels (wood, straw) has heating value in the range of 10-16 MJ/kg. The feed stock heating value should be more than 10 $MJkg^{-1}$.

8.2 Moisture Content

The heating value of the gas produced by any type of gasifier depends on the moisture content of the feedstock. High moisture contents cut the thermal efficiency since heat is used to drive off the water and consequently this energy is not used for the reduction reactions and for converting thermal energy into chemical bound energy in the gas. Therefore high moisture contents result in low gas heating values. The moisture content lower than 15% by weight is consider for trouble free and economical operation of the gasifier. In gasifiers high moisture contents resulted into low gas heating values as well as lower temperatures in the oxidation zone and this can lead to insufficient tar converting capability if the gas is use for engine applications.

8.3 Volatile Matter Content of The Fuel

Volatile matter and inherently bound water in the fuel is firm up in pyrolyis zone at the temperatures of 100-150°C forming a vapor consisting of water, tar, oils and gases. Fuel with elevated volatile matter content produces extra tar, causing problems to internal combustion engine. Volatile matters in the fuel decide the design of gasifier for removal of tar. In practice charcoal is considered as good biomass fuel because it contains least percentage of volatile matter (3-30 %) as compared to other biomass materials (crop residue : 63-80 %, Wood : 72-78 %). As a general rule if the fuel contains more than 10 percent volatile matter it should be used in downdraught gas producers.

8.4 Particle Size and Distribution

The fuel dimension affects the pressure drop across the gasifier and power that necessity be supplied to draw the air and gas through gasifier. Huge pressure drops will lead to reduction of the gas load in downdraft gasifier, consequential in low temperature and tar production. Very large sizes of particles give rise to lower reactivity of fuel, causing start-up problem and poor gas quality. Acceptable fuel sizes depend to certain extent on the design of gasifier. Normally wood gasifier work well on wood blocks and wood chips range 80 x 40 x 40 mm to 10 x 5 x 5 mm. For charcoal gasifier, charcoal with dimension range from 10 x 10 x 10 mm to 30 x 30 x 30 mm is fairly suitable.

8.5 Bulk Density of Fuel

Bulk density is defined as the weight per unit volume of loosely tipped fuel. Bulk density varies considerably with moisture content and particle size of fuel. Volume taken by stored fuel depend the bulk density of fuel as well as the manner in which fuel is piled. It is also recognized that bulk density has large impact on gas quality, as it influence the fuel residence time in the fire box, fuel velocity and gas flow rate.

8.6 Fuel Form

The shape in which fuel is fed to gasifier has an economical impact on gasification. Densification biomass has been practiced in the US for the past 40 years. Pelletize density all kinds of biomass and municipal waste into "Energy cubes". These cubes are available in cylindrical or cubic form and have a elevated density of 600-1000 kg/m^3. The specific volumetric content of cubes is more than the raw material from which they are made.

8.7 Ash Content of Fuel

A mineral content of fuel is which residue in oxidized form after combustion of fuel is called ash. In practice, ash also contains some unburned fuel. Ash content and ash composition have impact on easily operation of gasifier. Melting and agglomeration of ashes in reactor causes slagging and clinker formation. If no measures are taken, slagging or clinker formation leads to excessive tar formation or total blocking of reactor. In general, no slagging occurs when fuel consist of below 5% ash content. Ash content different from fuel to fuel. A wood chip contains 0.1% ash, while rice husk contains more amount of ash (16-23%).

8.8 Reactivity of Fuel

Reactivity determines the rate of decrease of carbon dioxide to carbon monoxide in the gasifier. Reactivity depends upon the type of fuel. It has found that coal is less reactive than wood and charcoal. There is connection between reactivity and the number of active places on the char surfaces. It is well recognized fact that reactivity of char surface can be enhanced through various processes including stream treatment (activated carbon) or treatment with lime and sodium carbonate. There are various elements which act as catalyst and influence the gasification process. Little quantities of potassium, sodium and zink can have huge influence on reactivity of the fuel.

9. IMPORTANT TERMINOLOGIES

Equivalence ratio: It is defined as the ratio of the actual fuel/air ratio to the stoichiometric fuel/air ratio. If the equivalence ratio is equal to one, the combustion is stoichiometric. If it is <1, the combustion is lean with excess air, and if it is >1, the combustion is affluent with incomplete combustion [59].

Turn down ratio: It is the ratio of maximum to minimum gas generation rates, at which it can be practically efficient, operated without drop in quality of gas.

Gasification Efficiency: It is the percentage energy of biomass converted into a cold producer gas (free from tar). It is expressed as:

$$\eta_g = \frac{\text{Calorific value of producer gas} \times \text{Total gas produced from 1 kg of biomass}}{\text{Average calorific value of biomass (per kg)}}$$

Specific Gasification Rate: It is the ratio of quantity (mass) of biomass consumed per unit time to cross-sectional area of gasifier. It is expressed as:

$$SGR = \frac{\text{Weight of dry biomass used per unit time (kg/h)}}{\text{Cross sectional are of the reactor (sq.m)}}$$

REFERENCES

1. Rathore NS, Panwar NL. Renewable energy sources for sustainable development, 1st ed. Himanshu Publication, Udaipur, Rajasthan; 2007.
2. Shay J. Elements of agricultural engineering, 4th ed. Standard Publisher, New Delhi; 2005.
3. Preeti H, Narnaware R, Surose S, Gaikwad. Current status and the future potentials of renewable energy in India - a review. International Journal of Advances in Science Engineering and Technology. 2015; 1.

4. Larisa S, Ramola A, Björn A, Claudia G. International Solid Waste Association. Austria: ISWA Report. 2015. https://doi.org/10.1177/0734242X17709910.
5. Nizami AS. Developing waste biorefinery in Makkah: A way forward to convert urban waste into renewable energy. [Applied Energy]. 2017; 186(2):189-196. https://doi.org/10.1016/j.apenergy.2016.04.116.
6. Hoornweg D, Bhada TP. What a waste: a global review of solid waste management. Urban development series knowledge papers; 2012. http://hdl.handle.net/10986/ 17388.
7. MNRE Govt. of India. Current status of bio energy in India, https://mnre.gov.in/bio-energy/current-status; 2020 [accessed on 22 August 2020].
8. National Policy for Management of Crop Residues (NPMCR). NPMCR report for management of crop residue in India, http://agricoop.nic.in/sites/default/files/ NPMCR_1. pdf; 2020 [accessed 23.08.2020].
9. Reed TB, Golden AD. Handbook of biomass downdraft gasifier engine systems. Washington: U.S. Government Printing Office; 1988.
10. Reed B, Graboski M. Biomass-to-methanol specialists workshop, Tamal Ton, CO, 3-5., SERI/CP-234-1590. Solar Energy Research Institute, Golden, Colo; 1982.
11. Rapagnà S., Kiennemann A., Foscolo PU., Steam-gasification of biomass in a fluidized-bed of olivine particles. Biomass Bioenergy. 2000; 19:187-197. https://doi.org /10.1016/S0961-9534(00)00031-3.
12. Parthasarathy P, Narayanan KS. Hydrogen production from steam gasification of biomass: Influence of process parameters on hydrogen yield. Renewable Energy. 2014, 66: 570-579. https://doi.org/10.1016/j.renene.2013.12.025.
13. All Power labs: Carbon Negative Power & Products. Gasification of biomass report, http://www.allpowerlabs.com/gasification-explained [accessed 20 March 2020].
14. Rai G.D., Non conventional energy sources. 4th ed. New Delhi: Khanna Publishers; 2002.
15. Kumar S, Kumar V, Sahu R. Fundamentals of agricultural engineering. New Delhi: Kalyani Publishers; 2016.
16. Pandey MM, Sirohi NPS, Singh KK, Ganesan S, Dhingra D. Handbook of agricultural engineering. 1st ed. New Delhi: Indian Council of Agricultural Research; 2013.
17. The Energy and Resources Institute India. Biomass gasifier for thermal and power applications, https://www.teriin.org/technology/ biomass-gasifier-for-thermal-and-power-applications; 2020 [accessed 9 July 2020]

18. Rahtore NS, Panwar NL, Kothari S. Biomass production and utilization technology. 1st ed. Udaipur: Himanshu Publication; 2007.
19. Biofuel Academy. Cross draft gasifier, https://www.biofuels academy.com; 2020 [accessed 9 July 2020]
20. Heidenreich S. Foscolo PU, New concepts in biomass gasification. Progress in energy and combustion science; 2015; 46:72-95. https://doi.org/10.1016/j.pecs.2014.06.002
21. Mc Kendry Eur. Ing. Peter, Energy Production from Biomass (Part 3): Gasification Technologies; 2020 [accessed 13 July 2020]
22. Fabry F, Rehmet C, Rohani V, Fulcheri L. Waste gasification by thermal plasma waste. Biomass Valorization. 2013; 4:421-439.https://doi.org/10.1007/s12649-013-9201-7.
23. Agon NM, Hrabovský O, Chumak M, Hlína V, Kopecký A, Maláni A, Bosmans L, Helsen S, Skoblja, Van Oost G, Plasma gasification of refuse derived fuel in a single-stage system using different gasifying agents. Waste Manage; 2016; 47: 246-255. https://doi.org/10.1016/j.wasman.2015.07.014
24. Tang L, Huang H, Hao H, Zhao K. Development of plasma pyrolysis/gasification systems for energy efficient and environmentally sound waste disposal. Journal of Electrostatics, 2013; 71:839-847. https://doi.org/10.1016/j.elstat.2013.06.007
25. Yang H, Yan R, Chen H, Zheng C, Lee, DH, Liang DT. In-depth investigation of biomass pyrolysis based on three major components: hemicellulose, cellulose and lignin. Energy Fuels. 2006; 20:388-393. https://doi.org/10.1021/ef0580117
26. Janajreh I, Raza S, Valmundsson AS., Plasma gasification process: Modeling, simulation and comparison with conventional air gasification. Energy Convers. Manage. 2013; 65:801-809. https://doi.org/10.1016/j.enconman.2012.03.010
27. Zhang L, Xu C, Champagne P. Overview of recent advances in thermo-chemical conversion of biomass. Energy Convers. Manage. 2010; 51:969-982. https:// doi.org/ 10.1016/ j.enconman.2009.11.038
28. Byrd AJ, Kumar S, Kong L, Ramsurn H, Gupta RB. Hydrogen production from catalytic gasification of switch grass biocrude in supercritical water. Int. J. Hydrogen Energy. 2011; 36: 3426-3433. https://doi.org/10.1016/j.ijhydene.2010.12.026
29. Guo L, Chen W, Cao H, Jin S, Guo, Zhang X. Hydrogen production by sewage sludge gasification in supercritical water with a fluidized bed reactor. Int. J. Hydrogen Energy. 2013; 38:12991-12999. https://doi.org/10.1016/j.ijhydene.2013.03.165

30. Kipçak E, Akgün M. Oxidative gasification of olive mill wastewater as a biomass source in supercritical water: Effects on gasification yield and biofuel composition. J. Supercrit. Fluids. 2012; 69: 57-63. https://doi.org/10.1016/j.supflu.2012.05.005

31. Miller A, Hendry D, Wilkinson N, Venkitasamy C, Jacoby W. Exploration of the gasification of Spirulina algae in supercritical water. Bioresour Technol. 2012; 119: 41-47. https://doi.org/10.1016/j.biortech.2012.05.005

32. Sricharoenchaikul V. Assessment of black liquor gasification in supercritical water. Bioresour Technol. 2009; 100:638-643. https:// doi.org/ 10.1016/j.biortech. 2008.07.011.

33. Basu P. Biomass gasification and pyrolysis. 1st ed. Boston: Academic Press; 2010.

34. Yanik J, Ebale S, Kruse A, Saglam M, Yüksel M. Biomass gasification in supercritical water: Part 1. Effect of the nature of biomass Fuel. 2007; 86:2410-2415. https://doi.org/10.1016/j.fuel.2007.01.025

35. Guo Y, Wang SZ, Xu DH, Gong YM, Ma HH, Tang XY. Review of catalytic supercritical water gasification for hydrogen production from biomass. Renewable Sustainable Energy Rev. 2010; 14:334-343.

36. Feng W, Van Der Kooi HJ, De Swaan Arons J. Biomass conversions in subcritical and supercritical water: driving force, phase equilibrium and thermodynamic analysis. Chem. Eng. Process. 2004; 43: 1459-1467. https://doi.org/10.1016/j.cep.2004.01.004

37. Florin NH, Harris AT. Enhanced hydrogen production from biomass with in situ carbon dioxide capture using calcium oxide sorbents. Chem. Eng. Sci. 2008: 63: 287-316. https://doi.org/10.1016/j.ces.2007.09.011

38. McKendry P. Energy production from biomass (part 3): gasification technologies. Bioresour. Technol. 2002; 83:55-63. https://doi.org/10.1016/S0960-8524(01)00120-1

39. Han L, Wang Q, Yang C, Yu, Fang M, Luo Z. Hydrogen production via CaO sorption enhanced anaerobic gasification of sawdust in a bubbling fluidized bed. Int. J. Hydrogen Energy. 2011; 36:4820-4829. https://doi.org/10.1016/j.ijhydene.2010. 12.086

40. Sutton D, Kelleher B, Ross JRH. Review of literature on catalysts for biomass gasification Fuel Process. Technol. 2001; 73: 55-173. https://doi.org/10.1016/S0378-3820 (01) 00208-9.

41. Xiao X, Le DD, Li L, Meng X, Cao J, Morishita K, Takarada T. Biomass steam gasification for hydrogen production: A systematic review. Biomass

Bioenergy. 2010; 34:1505-1512. http://doi-org-443.webvpn.fjmu.edu.cn/10.1007/978-3-319-07641-6_19.

42. Franco C, Pinto F, Gulyurtlu I, Cabrita I. The study of reactions influencing the biomass steam gasification process Fuel. 2003; 82: 835-842. https://doi.org/10.1016/S0016-2361(02)00313-7.

43. Florin NH, Harris AT. Hydrogen production from biomass coupled with carbon dioxide capture: The implications of thermodynamic equilibrium. Int. J. Hydrogen Energy. 2007; 32:4119-4134. https://doi.org/10.1016/j.ijhydene.2007.06.016

44. Rapagná SH, Provendier C, Petit A, Kiennemann, Foscolo PU. Development of catalysts suitable for hydrogen or syn-gas production from biomass gasification. Biomass Bioenergy. 2002; 22:377-388. https://doi.org/10.1016/S0961-9534(02)00011-9

45. Ross D, Noda R, Horio M, Kosminski A, Ashman P, Mullinger P. Axial gas profiles in a bubbling fluidised bed biomass gasifier. Fuel. 2007; 86:1417-1429. https://doi.org/10.1016/j.fuel.2006.11.028

46. Weerachanchai P, Horio M, Tangsathitkulchai C. Effects of gasifying conditions and bed materials on fluidized bed steam gasification of wood biomass. Bioresour. Technol. 2009; 100:1419-1427. https://doi.org/10.1016/j.biortech.2008.08.002

47. Chen S, Wang D, Xue Z, Sun X, Xiang W. Calcium looping gasification for high-concentration hydrogen production with CO_2 capture in a novel compact fluidized bed: Simulation and operation requirements. Int. J. Hydrogen Energy. 2011; 36: 4887-4899. https://doi.org/10.1016/j.ijhydene.2010.12.130

48. Balat M, Balat M, Kirtay E, Balat H. Main routes for the thermo-conversion of biomass into fuels and chemicals. Part 2: Gasification systems. Energy Convers. Manage. 2009; 50:3158-3168. https://doi.org/10.1016/j.enconman.2009.08.013

49. Tanksale AJ, Beltramini N, Lu G. M. A review of catalytic hydrogen production processes from biomass. Renewable Sustainable Energy Rev. 2010; 14:166-182. https://doi.org/10.1016/j.rser.2009.08.010.

50. Acharya B, Dutta A, Basu P. Chemical-looping gasification of biomass for hydrogen-enriched gas production with in-process carbon dioxide capture. Energy Fuels. 2009; 23:5077-5083. https://doi.org/10.1021/ef9003889.

51. Acharya B, Dutta A, Basu P. An investigation into steam gasification of biomass for hydrogen enriched gas production in presence of CaO. Int. J. Hydrogen Energy. 2010; 35:1582-1589. https://doi.org/10.1016/j.ijhydene.2009.11.109.

52. Guoxin H, Hao H. Hydrogen rich fuel gas production by gasification of wet biomass using a CO_2 sorbent. Biomass Bioenergy. 2009; 33:899-906. https://doi.org/10.1016/j.biombioe.2009.02.006.

53. Hanaoka T, Yoshida T, Fujimoto S, Kamei K, Harada M, Suzuki Y, Hatano H, Yokoyama SY, Minowa T. Hydrogen production from woody biomass by steam gasification using a CO_2 sorbent. Biomass Bioenergy. 2005; 28: 63-68. https://doi.org/10.1016/j.biombioe.2004.03.009.

54. Howaniec N, Smoliñski A. Steam gasification of energy crops of high cultivation potential in Poland to hydrogen-rich gas. Int. J. Hydrogen Energy.2011; 36: 2038-2043. https://doi.org/10.1016/j.ijhydene.2010.11.049.

55. Wei L, Xu S, Liu J, Liu C, Liu S. Hydrogen production in steam gasification of biomass with Cao as a CO_2 absorbent. Energy Fuels. 2008; 22:1997-2004. https://doi.org/10.1021/ef700744a.

56. Rajvanshi A. Biomass gasification. Alternative Energy in Agriculture. 1986; 2:82-102.

57. National Energy Technology Laboratory USA. Energy from gasification, http://www.netl.doe.gov/research/coal/energy-systems/gasification/gasifipedia/wabash/;2000.

58. Manovic V, Anthony E. J. Lime-based sorbents for high-temperature CO_2 capture—a review of sorbent modification methods. Int. J. Environ. Res. Public Health. 2010; 7:3129-3140. https://doi.org/10.3390/ijerph7083129

59. Bioenergy consult. Gasification terminology, https://www.bioenergyconsult.com/biomass-gasification/; 2020 [accessed 07 July 2020].

Bioenergy Engineering, Pages: 113–130
Edited by: Mahendra S. Seveda, Pradip D. Narale and Sudhir N. Kharpude
Copyright © 2021, Narendra Publishing House, Delhi, India

CHAPTER - 6

TORREFACTION OF BIOMASS

Km. Sheetal Banga[1*], Sunil Kumar[2] and Raveena Kargwal[2]

[1]*Janta Vedic College, Baraut*
Chaudhary Charan Singh University Meerut-250611, Uttar Pradesh, India
[2]*College of Agricultural Engineering and Technology*
Chaudhary Charan Singh Haryana Agricultural University,
Hisar-125004, Haryana, India
*Corresponding Author

Biomass torrefaction has gained prevalent consideration due to its aids as a stand-alone process to improve biomass properties to be at par or similar to those for coal in electricity generation or as a pretreatment step before pyrolysis and gasification processes. It has also establish application in other processes like steel production where it is aiming to replace coal or work alongside coal. Torrefaction of biomass can be described as a mild form of pyrolysis at temperatures typically ranging between 200 and 300°C in an inert and reduced environment. The main product of the process, torrefied biomass determines the efficiency and how it can be applied to other technologies. To date, biomass torrefaction is for co-firing with coal for energy generation and as a pretreatment step for pyrolysis and gasification. Due to varying types of biomass in different countries, the technology has not yet reached its full potential, but the hope is it will with calls for use of renewable sources of energy.

1. INTRODUCTION

Bioenergy derived from biomass is a renewable source of energy which can be utilized as an alternative to non-renewable sources of energy like coal for energy generation (and electricity production). However, biomass as provided by nature has less energy density, more moisture and volatiles when compared to coal. As a result of this, the biomass needs a pretreatment process so as to improve its properties before it can be used together with or as a replacement for coal.

Torrefaction, a thermochemical process, is regarded as a simple and effective method to transform the biomass properties to become almost at par with those of coal The word of torrefaction (French word for "roasting") has been used in a mass of industries for tea and coffee making, but only in recent time, it has caught the attention of power industries for the production of a coal substitute from biomass. Torrefaction is often called a pretreatment process as it prepares biomass for further use instead of direct use in its raw form. It is also known as slow pyrolysis, mild pyrolysis, wood cooking or high-temperature drying. Torrefied biomass have many uses as given below:

- Co-firing biomass with coal in large coal-fired power plant boilers.
- Use as fuel in decentralized or residential heating system.
- Use as a convenient fuel for gasification.
- Potential feedstock for chemical industries.
- Substitute for coke in blast furnace for reduction in carbon footprint.

2. WHAT IS TORREFACTION?

Torrefaction stage is key to the whole process as the bulk of the depolymerization of the biomass takes place in this stage. A certain amount of time is needed to allow the desired degree of depolymerization of the biomass to occur. The degree of torrefaction depends on the torrefaction temperature as well as on the time the biomass is subjected to torrefaction. This time is also called reactor residence time or torrefaction time. The torrefaction time should be measured from the instant the biomass reaches the temperature for the onset of torrefaction (200°C) because the degradation of biomass below this temperature is negligible. It is defined as:

It is a thermo chemical process in an inert or limited oxygen environment where bio-mass is slowly heated to within a temperature range of 200–300°C and retained there for a stipulated time such that it results in near complete degradation of its hemicellulose content while maximizing mass and energy yield of solid product.

The torrefaction process is mildly exothermic [11] over the temperature range of 250–300°C. So, the torrefaction stage should require very little energy, but in practice it could require some heat to make up for the heat loss from the torrefaction section of the reactor. Due to the compression strength used in the torrefaction process, the pellets do not disintegrate easily during the handling and storage phases, as they have a 1.5 to 2 times impact load. Simultaneously, all biological activity is stopped, reducing the risk of fire and stopping biological decomposition like rotting [10].

3. TORREFACTION PROCESS TECHNIQUE

It is a thermochemical process involving the interface of drying and incomplete pyrolysis. The different parameters that influence the torrefaction process are (a) reaction temperature, (b) heating rate, (c) absence of oxygen, (d) residence time, (e) ambient pressure, (d) flexible feedstock, (e) feedstock moisture, and (f) feedstock particle size.

Biomass feedstock is typically predried to 10% or less moisture content prior to torrefaction. Particle size plays an important role in torrefaction which influences the reaction mechanisms, kinetics, and duration of the process, given a specific heating rate. The chemical reactions that occur when reactive intermediates are trapped in a thick matrix differ from the situations in which products can escape and be swept away in a gas stream. The duration of the process is basically adjusted to produce friable, hydrophobic, and energy-rich enhanced biomass fuel.

3.1 Reaction Temperature

Torrefaction temperature has the greatest influence on torrefaction as the degree of thermal degradation of biomass depends primarily on the tem-perature. Higher temperature gives lower mass and energy yields but higher energy density. The fraction of fixed carbon in a sample increases while that of hydrogen and oxygen decreases as the torre-faction temperature increases. Typical temperature range for this process is between 200 and 300°C. Torrefaction above this tempera-ture would result in extensive devolatilization and carbonization of the polymers both of which are undesirable for torrefaction. Also, the loss of lignin in biomass is very high above 300°C. This loss could make it dif-ficult to form pellets from torrefied products. Furthermore, fast thermal crack-ing of cellulose causing tar formation starts at temperature 300–320°C [2]. These reasons fix the upper limit of torrefaction tempera-ture as 300°C. A major motivation of torrefaction is to make the biomass lose its fibrous nature such that it is easily grindable, while it is still possible to form into pellets without binders. Such requirements limit the torrefaction tempera-ture range to 200–300°C. It is founded that the optimum condition near 280 °C and 60 min of residence time in the torrefaction of eucalyptus [12]. Effect of different temperature on degradation of lignocellulosic components is enlisted below in Table 6.1.

Table 6.1: Degradation of Lignocellulosic Components at Different Temperature [8]

Temperature Range (°C)	Hemicellulose	Cellulose	Lignin
Overall degradation temperature range	180-300 °C	275-355 °C	250–500°C
25-105	No effect	No effect	No effect
105-150	No effect	No effect	Softening occurs at T> 130 °C but no degradation
160-180	Degradation starts due to devolatilization and depolymerization reaction releasing mostly H_2O and small amount of CO_2	No effect	No effect
180-200	Endothermic reactions	N/A	N/A
200	Light torrefaction		
200-250	Degradation continues. Colour changes for biomass is noticeable. Volatiles like acetic acid, methanol, CO and CO_2 are formed	Colour changes for biomass is noticeable.	Colour changes for biomass is noticeable.
250	Mild Torrefaction		
200-270	Partly endothermic reactions	Partly endothermic reactions	Partly endothermic reactions
250-300	Total degradation forming char and release of CO, CO_2 and H_2O	Degradation start sat~275°C releasing H_2O and forming an hydrous cellulose and levoglucosan polymer	Degradation starts at~250°C. At 280°C degradation gives phenolsdue to cleavage of ether bonds
290	Severe torrefaction	Depolymerization of cellulose	Depolymerization of lignin
250-300	Depolymerization of remaining hemicelluloses		
270-300	Exothermic reaction	Exothermic reactions	Exothermic reactions
330-370		Total degradation forming char	Total degradation forming char

3.1.1 Core Temperature Rise

Torrefaction often involves coarse biomass particles few millimeter or centime-ter in size, where the reaction takes place mainly within its interior. Temperature inside the particle is thus more important than the one on its surface for the decomposition. Below 270°C, torrefaction process is mildly endothermic, but it is mildly exothermic above 280°C possibly due to exothermic breakdown of sugars at higher temperature [9]. The magnitude of heat of reaction is however small. In any case, the torrefaction heat is transported from an external heat source first to the biomass particle's outer surface. Thereafter, the heat travels to its interior by conduction and pore convection. Thus one would expect temperature gradient between the torrefac-tion reactor, particle surface, and its interior (core). Typical thermal treatment process variables (mass and energy yields at different temperature regimes) are enlisted in Table 6.2.

3.2 Heating Rate

Slow heating rate is an important characteristic of torrefaction. Unlike in pyrol-ysis, the heating rate in torrefaction must be sufficiently slow to allow maximi-zation of solid yield of the process. Typically the heating rate of torrefaction is less than 50°C/min [3]. A higher heating rate (>10 min) would increase liquid yield at the expense of solid products as is done for pyrolysis.

The thermal decomposition of biomass occurs via a series of chemical reac-tions coupled with heat and mass transfer. Within the temperature range of 100–260°C, hemicellulose is chemically most active, but its major degradation starts above 200°C. Cellulose degrades at still higher temperature (>275°C), but its major degradation occurs within a narrow temperature band of 270–350°C. Lignin degrades gradually over the temperature range of 250–500°C, though it starts softening in the temperature range of 80–90°.

3.3 Residence Time

The time duration of torrefaction at the torrefaction temperature is referred to as "residence time," and it exert an important influence on the thermal degradation of biomass. Longer residence time gives lower mass yield and higher energy density.

3.4 Biomass Type

The biomass type is another parameter that could influence torrefaction. As hemi-cellulose degrades most within the torrefaction temperature range (200–

Table 6.2: Thermal Treatment Process Variables for Different Temperature Regimes [14]

Temperature (°C)	Time (min)	Process reactions	Heating rate (°C/min)	Drying environment and pressure	Mass yield (%)	Energy yield (%)
50–150	30–120	Nonreactive drying (moisture removal and structural changes	<50	Air and ambient pressure	~90–95	Not significant
150-200	30–120	Reactive drying (moisture removal and structural damage due to cell wall collapse)	<50	Air and ambient pressure	~90	Needs to be researched
200-300	<30	Destructive drying, Devolatilization and carbonization of hemicellulose, Depolymerization and devolatilization/softening of lignin, Depolymerization and devolatilization of cellulose	<50	Inert environment and ambient pressure	~70	~90

300°C), one would expect a higher mass loss in a biomass with high hemicellulose con-tent. It is interesting that, when torrefied under identical conditions, a hardwood and softwood with similar hemicellulose content could show very different mass yields [7,11]. Torrefaction of hardwood gives low mass yield because xylan, the active component of hemicellulose, constitutes 80%–90% of the total hemicellulose in hardwood (deciduous), while in softwood (coniferous), it constitutes only 15%–30% [2]. Various studies conducted by many researchers for the torrefaction of different types of biomass are enlisted in Table 6.3.

Table 6.3: Different Methods and Various Biomass at Different Conditions [8]

Types of biomass	Conditions				
	T (°C)	P (Atm)	t (mins)	HR (°C/min)	O_2 (%)
Beech, Willow, Larch, Straw	230-300	1	92	10-100	0
Beech, Willow	220-300	1	10-60	10-20	0
Larch wood	230-290	1	10-50	10-20	0
Straw	200	1	30	10-20	0
Rice straw, Rape stalk	200-300	0.019	30	30-45	0
Rice straw, Cotton gin waste	260	1	15-60	-	0
Wheat straw	200-315	1	15-180	-	0
Sawdust	230-290	1	20-30	10	0
Bagasses, Road side Grass, Popular Straw	240-280	1	30	5	0
RDF/SRF, Grass Seed hay, Spruce chips	240-300	1	30	5	0
Pine chips, Trockenstabilat	260-300	1	30	5	0
Beech	280	1	30	5	0
LeucaenaLeucaephala	200-275	1-49	20-120	10	0
Microalgae, Microalgal residue	200-350	1	30-90	30-50	0
Corn stover	200-300	1	10-30	-	0
LeucaenaLeucaephala	200-250	1	30	-	0
Pine chips, spruce chips, Fir, SPF, Pine bark	280	1	52	-	0
SPF shavings	240-340	1	60	-	0
Fir Sawdust pellets	240-310	1	8-22	-	0
Palm Kernal Shell	200-350	1	10-60	-	0
Saw dust	200-300		166.8	-	0
Oil palm fiber, Coconut fiber, eucalyptus, cryptomeriajaponics	300		60	-	0-21

[Table Contd.

Contd. Table]

Types of biomass	Conditions				
	T (°C)	P (Atm)	t (mins)	HR (°C/min)	O_2 (%)
Palm Kernal Shell	200-300		30	-	0
Poplar	220-300	1-5.9	15-35	-	0-21
Beech	220-300	1	25-175	10	0
Palm Kernal Shell 15%	250	1	30	10	0-15 CO_2: 9-
Bamboo	200-300	1	60	5	0
waste	150-600	1	60	10-30	0
Wood, Cow dung, Maize corn cobs	200-300	1	40	10	0
Marula trees blue gum wood	200-300	1	20-60	5-15	0-20

3.5 Ambience

Another important aspect of torrefaction's definition is oxygen concentration in the reactor. Studies on the effect of oxygen concentration on torrefaction suggest that it is not essential to have oxygen-free environment for torrefaction. Presence of a modest amount of oxy-gen can be tolerated and may even have a beneficial effect on the torrefaction.

4. MECHANISM OF TORREFACTION

Torrefaction involves devolatilization and carbonization of the biomass poly-mers. Complete degradation of all these polymers do not necessarily take place within the narrow (200–300°C) temperature range of torrefaction. Different polymers degrade in different temperature ranges. Some qualitative values taken from [11] are given below.

Hemicellulose: 225–300°C

Cellulose: 305–375°C

Lignin: 250–500°C

The thermochemical changes in biomass during torrefaction may be divided into five regimes following the description of (3):

i. **Regime A (50–120°C):** This is a *non-reactive* drying regime where there is a loss in physical moisture in biomass but no change in its chemical composi-tion. The biomass shrinks but may regain its structure if rewetted [14]. Upper temperature limit of the regime is higher for cellulose.

ii. ***Regime B (120–150°C):*** This regime is separated out only in the case of lignin that undergoes softening.

iii. ***Regime C (150–200°C):*** This is called *"reactive drying"* regime that results in structural deformity of the biomass. This deformation cannot be regained upon cooling or wetting. This stage initiates breakage of hydrogen and carbon bonds and depolymerization of hemicellulose. This stage produces shortened polymers that condense within solid structures.

iv. ***Regime D (200–250°C):*** This regime along with regime (E) constitutes the torrefaction zone for hemicellulose. This regime is characterized by limited devolatilization and carbonization of solids structure formed in regime (C). It results in the breakdown of most inter- and intra-molecular hydrogen, CC, and CO bonds forming condensable liquids and non-condensable gases [14].

v. ***Regime E (250–300°C):*** This is the highest part of the torrefaction process. Extensive decomposition of hemicellulose into volatiles and solid products takes place. Lignin and cellulose, however, undergo only a limited amount of devolatilization and carbonization. Biomass cell structure is completely destroyed in this regime making it brittle and non-fibrous.

5. TORREFACTION PRODUCTS

During torrefaction, three different products are produced:

a. brown to black uniform solid biomass, which is used for bioenergy applications,

b. condensable volatile organic compounds comprising water, acetic acid, aldehydes, alcohols, and ketones, and

c. noncondensable gases like CO_2, CO, and small amounts of methane.

Release of these condensable and non-condensable products results in changes in the physical, chemical, and storage properties of biomass. Several studies have also investigated the physical properties and chemical composition of the liquids and gases released during torrefaction. An overview of the torrefaction products, based on their states at room temperature, which can be solid, liquid, or gas[3]. The solid phase consists of a chaotic structure of the original sugars and reaction products. The gas phase includes gases that are considered permanent gases, and light aromatic components such as benzene and toluene. The condensables, or liquids, can be further divided into four subgroups: (1) reaction water produced from thermal decomposition, (2) freely bound water that has been released through evaporation, (3) organics (in liquid form), which consist of organics produced during devolatilization and carbonization, and (4) lipids, which contain compounds such as waxes and fatty acids. Condensable and noncondensable products are emitted from the biomass based on heating rate, torrefaction temperature and time, and biomass composition. The emission profiles of these products greatly depend on the moisture content in the biomass.

6. SOLID TORREFIED BIOMASS PROPERTIES

Torrefaction of biomass significantly changes its physical and chemical properties like moisture content, density, grindability, pelletability, hydrophobicity, calfonic value, proximate and ultimate composition, and storage behaviors in terms of off-gassing, spontaneous combustion, and self-heating.

6.1 Physical Properties

6.1.1 Moisture Content

Normally, feedstock moisture content ranges from 10–50%, but because torrefaction is a deep drying process, moisture content is reduced to 1–3% on a weight basis, depending on the torrefaction conditions. Typically, torrefaction achieves an equilibrium moisture content of 3% and a reduction of mass by 20–30% (primarily by release of water, carbon oxides, and volatiles), while retaining 80–90% of the wood's original energy content. Reduction in moisture during torrefaction provides three main benefits: (1) reduced moisture level for the conversion process, (2) reduced transportation costs associated with moving unwanted water and (3) the prevention of biomass decomposition and moisture absorption during storage and transportation.

6.1.2 Bulk and Energy Density

Mass loss in the form of solids, liquids, and gases during torrefaction cause the biomass to become more porous. This results in significantly reduced volumetric density, typically between 180 and 300 kg/m^3, depending on initial biomass density and torrefaction conditions.

6.1.3 Grindability

Biomass is highly fibrous and tenacious in nature; fibers form links between particles and make handling the raw ground samples difficult. During torrefaction, the biomass loses its tenacious nature, which is mainly associated with the breakdown of the hemicellulose matrix and depolymerization of the cellulose, resulting in decreased fiber length. Particle size distribution, sphericity, and particle surface area Particle-size distribution curves, sphericity, and surface area are important parameters for understanding flowability and combustion behavior during cofiring. Many researchers observed that ground, torrefied biomass produced narrower, more uniform particle sizes compared to untreated biomass due to its brittle nature, which is similar to coal.

Torrefaction also significantly influences the sphericity and particle surface area. Phanphanich and Sudhagar's results also indicated that sphericity and particle surface area increased as the torrefaction temperature was increased up to 300°C [5,6].

6.1.4 Pelletability

Variability in feedstock quality due to differences in the types of raw materials, tree species, climatic and seasonal variations, storage conditions, and time significantly influence the quality of biopellets. Torrefying the biomass before pelletization, however, produces uniform feedstock with consistent quality.

6.2 Chemical Compositional Changes

Besides improving physical attributes, torrefaction also results in significant changes in proximate and ultimate composition of the biomass and makes it more suitable for fuel applications.

6.2.1 Calorific Value

Biomass loses relatively more oxygen and hydrogen than carbon during torrefaction, which increases the calorific value of the product. The calorific value of torrefiedwood can reach a calorific value close to coal and is very dry (moisture content lower than 5%). It contains less ash than coal (0.7 to 5% db, compared to 10 to 20% db for coal), and has a higher reactivity, largely by the virtue of high amounts of volatile matter (55 – 65% db compared to 10 – 12% db for coal). The net CV of torrefied biomass is 18–23 MJ/kg (lower heating value [LHV], dry) or 20–24 MJ/kg (higher heating value [HHV], dry).

7. STORAGE ASPECTS OF TORREFIED BIOMASS

7.1 Off-gassing

Storage issues like off-gassing and self-heating may also be insignificant in torrefied biomass, as most of the solid, liquid, and gaseous products that are chemically and microbiologically active are removed during the torrefaction process. Some studies on wood pellets concluded that high storage temperatures of 50°C can result in high CO and CO_2 emissions, and the concentrations of these off-gases can reach up to 1.5% and 6% for a 60-day storage period [14]. Comparison of different properties of torrefied pellets with coal are enlisted in Table 6.4.

Table 6.4: Comparison of Different Properties: Torrefied Pellets v/s Coal [1]

Parameters	Wood	Wood pellet	Torrefied pellet	Coal
Moisture content (wt. %)	30–40	7–10	1–5	10–15
Calorific value (MJ/kg)	9–12	15–16	20–24	23–28
Volatiles (% db)	70–75	70–75	55–65	15–30
Fixed carbon (% db)	20–25	20–25	28–35	50–55
Bulk density (kg/m^3)	200–250	550–750	750–850	800–850
Volumetric energy density (GJ/m^3)	2.0–3.0	7.5–10.4	15.0–18.7	18.4–23.8
Dust explosibility	Average	Limited	Limited	Limited
Hydroscopic properties	Hydrophilic	Hydrophilic	Hydrophobic	Hydrophobic
Biological degradation	Yes	Yes	No	No
Milling requirements	Special	Special	Classic	Classic
Handling properties	Special	Easy	Easy	Easy
Transport cost	High	Average	Low	Low

7.2 Hydrophobicity

In general, the uptake of water by raw biomass is due to the presence of OH groups. Torrefaction produces a hydrophobic product by destroying OH groups and causing the biomass to lose the capacity to form hydrogen bonds [4]. Due to these chemical rearrangement reactions, nonpolar unsaturated structures are formed, which preserve the biomass for a long time without biological degradation, similar to coal.

8. APPLICATIONS WITH TORREFIED BIOMASS

Torrefaction allows products with great uniformity. From the same process we can manufacture fuels for different purposes. Some applications will be described as follows:

i. **Gasification:** torrefaction is recommended as a pretreatment of the biomass before gasification, as it decreases the mechanical properties, as a fibrous structure facilitating the gasification process. Torrefied biomass needs less energy for milling.

ii. **Industrial and domestic fuel:** The fact of having low emissions of fumes during combustion, in addition to being able to be stored for long periods, are factors that favor the domestic use of the wood. Industrially toasted biomass can be used in large scale for the production of electricity in the firing in boilers for the production of steam.

iii. **Another alternative use** would be the co-combustion with mineral coal, as it would provide environmental benefits by reducing sulfur dioxide emissions.

iv. **Reduction:** The high fixed carbon content of torrefied wood provides potential to be applied as a reducer in the metallurgical industry. Experiments carried out in a furnace for the production of silicon, which requires high mechanical strength reducers, where toasted wood is more efficient than the traditionally used mixture of charcoal and torrefied wood.

9. CLASSIFICATION OF REACTORS USED IN TORREFACTION

With diverse technologies, different types of reactors are used in torrefaction. From this variety, two main groups of reactors can be classified as reactors with direct and indirect heating (Fig. 6.1). Indirect heating reactors can be further

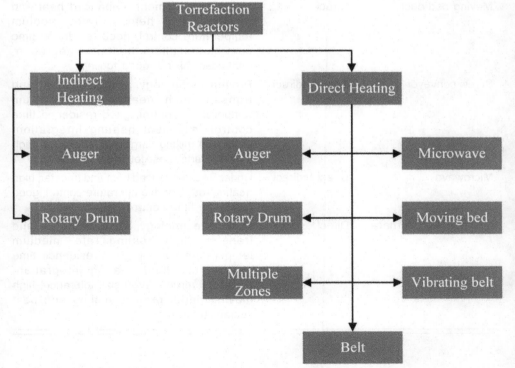

Figure 6.1: Classification of Torrefaction Reactors [12]

classified as: auger and rotary type. According to oxygen content in the heating medium direct heating group may be divided into several subgroups: (1) the reactors in which the heating medium does not contain oxygen and (2) reactors wherein the heating medium contains a small amount of oxygen and other types.

Table 6.5: Comparison of Different Torrefaction Potential Technologies [1]

Torrefaction Technology	Mode of Heating	Status Criteria
Rotary drumreactor	Direct	Proven technology, minimum heat transfer, high heating rate, medium temperature control, good residence time control, excellent heating integration, enhanced mixing, large size tolerance, high moving parts, good fouling,, little scaling problem
Fluidized bedreactor	Direct	Proven technology, enhanced heat transfer, high heating rate, medium temperature control, medium residence time control, excellent scalability, excellent heating integration, excellent uniform heating materials, enhanced mixing
Moving bedreactor	Direct	Under development, enhanced heat and transfer, high heating rate, medium temperature control, good residence time control, excellent heating integration, enhanced mixing, good fouling
Screw conveyor	Direct Indirect	Proven technology, enhanced heat and transfer, high heating rate, medium temperature control, good residence time control, excellent heating integration, enhanced mixing, large size tolerance, high moving parts, best fouling and scaling
Microwave	Direct Indirect	Under R&D, enhanced heat and transfer, high heating rate, good temperature control, good residence time control
Multiple hearthfurnace	Direct	Proven technology, enhanced heat and transfer, high heating rate, medium temperature control, good residence time control, excellent heating integration, enhanced mixing, large size tolerance, high moving parts, perfect scaling and best scalability

10. ADVANTAGES OF TORREFIED BIOMASS

Torrefaction has a great effect on the physical and chemical properties of biomass, so this technology brings some advantages, such as:

1. It allows for the conversion of biodegradable biomass into a hydrophobic product which is not prone to biological decomposition. This conversion allows for long-term preservation of the processed biomass that would otherwise suffer from degradation in a storage environment that is exposed to the weather.
2. Agricultural waste does not have favorable heating properties, such as very high humidity and low energy density. As it removes moisture from the agricultural residues and increases the energy density of the residues, due to it torrefaction process is a promising method for the pretreatment of residues.
3. The grind ability is facilitated and, as a consequence, the energy consumption for milling is three to seven times lower than that of the raw material, which will not undergo the torrefaction process.
4. Due to increment in porosity, torrefied biomass becomes more reactive during combustion and gasification.
5. The torrefaction process makes the transport and storage logistics of biomass more efficient, since the cost with transportation, storage, and transshipments are mainly based on the volume of the material. It reduces the volume of the biomass transported as an economic propellant prior to transport.
6. It a less energy intensive conversion process. It opens the door for utilizing less expensive equipment and materials, which aids in producing more economical conversion systems.

11. CONCLUSIONS

Potential of torrefaction of biomass materials is growing as it improves the quality of both herbaceous and woody materials. Some points are discussed below which meet the needs of energy providers and has discussed options for improving biomass resources for use in power and liquid-fuel production. These areas provide a brief understanding of the potential use of torrefaction as a means of improving the physical, chemical, and rheological characteristics of biomass materials.

i. Torrefied biomass, in general, defines a group of products resulting from the partially controlled and isothermal pyrolysis of biomass occurring at the 200–300°C temperature range.

ii. The most common torrefaction reactions include (a) devolatilization and carbonization of hemicelluloses, and (b) depolymerization and devolatilization of lignin and cellulose.

iii. Torrefaction of the biomass helps in developing a uniform feedstock with minimum variability in moisture content.

iv. Torrefaction of biomass improves (a) energy density, grind ability, and pellet ability index ratings, (b) ultimate and proximate composition by increasing the carbon content and CV and decreasing the moisture and oxygen content, and (c) biochemical composition by decomposing the hemicelluloses and softening the lignin, which results in better binding during pelletization.

v. Ground torrefied biomass has improved sphericity, particle surface area, and particle size distribution.

vi. Densification of torrefied material reduces specific energy consumption and increases throughput by about a factor of 2 compared to raw or untreated biomass.

vii. During torrefaction the biomass loses most of the low energy content of the material, like (a) solids, which include original sugar structures and other newly formed polymeric structures, and (b) liquids, which include condensable, like water, organics, and lipids, and (c) gases, which include H_2, CO, CO_2 and CH_4, C_xH_y, toluene, and benzene.

viii. Torrefaction preserves the biomass for a long time without biological degradation due to the chemical rearrangement reactions and formation of nonpolar unsaturated structures.

ix. Torrefied biomass has been successfully used as an upgraded solid fuel in electric power plants and gasification plants.

Not all aspects of torrefaction and its influence on other processing operations have been reconnoitered. Prospects for future research can include:

i. Optimizing torrefaction processes based on activation energies required to degrade the cellulose, hemicelluloses, and lignin.

ii. Understanding the torrefaction process at a molecular level by identifying different functional groups and energies associated with breaking the bonds.

iii. Understanding the spontaneous torrefaction process reactions using Fourier transform infrared (FTIR) and Raman spectroscopy.

iv. Understanding the severity of the torrefaction process based on color changes using the Hunter colorimeter.

v. Studies on thermogravimetrics to establish the weight-loss kinetics.

vi. Studies on microstructural changes in biomass at different temperature regimes.

vii. Testing integrated processes such as torrefaction and densification.

viii. Calculating energies associated with condensable and non-condensable products produced during torrefaction and the ability to reuse them to improve the overall process efficiency.

ix. Understanding the storage behavior of torrefied biomass in terms of off-gassing.

REFERENCES

1. Acharya B, Sule I, Dutta, A. A review on advances of torrefaction technologies for biomass processing. Biomass Conv Bioref. 2012; 2: 349–369.

2. Barbanera M, Muguerza IF. Effect of the temperature on the spent coffee grounds torrefaction process in a continuous pilot-scale reactor. Fuel. 2020; 262: e116493.

3. Bergman PCA, Kiel JHA. Torrefaction for biomass upgrading. In: Proceedings of the 14th European Biomass Conference and Exhibition. Paris, France; 2005.

4. Couhert C, Salvador S, Commandre, JM. Impact of torrefaction on syngas production from wood. Fuel. 2009; 88(11): 2286-2290.

5. Almeida G, Brito JO, Perré P. Alterations in energy properties of eucalyptus wood and bark subjected to torrefaction: the potential of mass loss as a synthetic indicator. Bioresource Technology. 2010; 101(24): 9778-9784.

6. Kambo HS, Dutta A. Strength, storage, and combustion characteristics of densified lignocellulosic biomass produced via torrefaction and hydrothermal carbonization. Applied Energy. 2014; 135: 182-191.

7. Mafu LD, Neomagus HWJP, Everson RC, Carrier M, Strydom CA, Bunt JR. Structural and chemical modifications of typical South African biomasses during torrefaction. Bioresour. Technol. 2016; 202, 192-197.

8. Mamvura TA, Danha G. Biomass torrefaction as an emerging technology to aid in energy production. Heliyon. 2020; 6(3): e03531.

9. Medic D, Darr M, Shah A, Rahn S. The effects of particle size, different corn stover components, and gas residence time on torrefaction of corn stover. Energies. 2012; 5(4): 1199-1214.

10. Kongkeaw N, Patumsawad S. Thermal upgrading of biomass as a fuel by torrefaction, in: Second International Conference on Environmental Engineering and Applications, IACSIT Press, Singapore; 2011.
11. Prins MJ, Ptasinski KJ, Janssen FJJG. Torrefaction of wood: Part 2. Analysis of products. J Anal Appl Pyrol. 2006; 77(1): 35–40
12. Stêpieñ P, Pulka, J, Biaowiec A. Organic waste torrefaction–a review: reactor systems, and the biochar properties. Pyrolysis. 2017; 37.
13. Singh RK, Sarkar A, Chakraborty JP. Effect of torrefaction on the physicochemical properties of eucalyptus derived biofuels: estimation of kinetic parameters and optimizing torrefaction using response surface methodology (RSM). Energy. 2020; 31:117369.
14. Tumuluru JS, Sokhansanj S, Wright CT, Hess JR, Boardman RD. A review on biomass torrefaction process and product properties (No. INL/CON-11-22634). Idaho National Laboratory (INL); 2011.

CHAPTER - 7

ALGAL BIOMASS: A PROMISING SOURCE FOR FUTURE BIOENERGY PRODUCTION

Subodh Kumar*, Adya Isha, Ram Chandra, Anushree Malik and Virendra K. Vijay

Centre for Rural Development and Technology
Indian Institute of Technology Delhi, Hauz Khas, New Delhi-110016, India
*Corresponding Author

Biomass to bioenergy is emerging as the renewable sustainable alternative of fossil fuel power production. Agricultural residues, organic fraction of municipal solid waste, kitchen waste, and algal biomass are the major source for biofuel production such as bioethanol, biodiesel, and biogas, utilizing various technologies. Algal biomass is emerging as a potential aquatic source for biofuel production due to its high productivity and ability to grow in non-arable land. However, algal biomass is not utilized commercially as the source of biofuel production due to some techno-economic limitations such as costly and energy-intensive harvesting methods, low energy density of algal biomass, and low total solid content of algal biomass. Anaerobic digestion (AD) is the favourable biochemical conversion technology for converting the biomass into biofuel (biogas) that can tackle the challenge of the low total solid content of algal biomass with more energy-efficient way. However, the practical achievements of biogas yield from algal biomass are significantly below the calculated theoretical yields. New processing technologies are required to make the algal biomass more cost-effective and energy-efficient for biogas production. This chapter discusses the characteristics of algal (macroalgae and microalgae) biomass, potential of algal biomass for biofuel (biogas, bioethanol, biodiesel) production, challenges in cultivation, technologies to enhance the biogas yield, life cycle assessment and strategies to make algal biomass to bioenergy production a commercially feasible option.

1. INTRODUCTION

Renewable and sustainable sources are the inevitable need in the present scenario of climatic changing conditions, declining reserves of fossil fuel, due to the expedition of the human population and industrialization. In this way, the ample amount of organic residues such as agricultural residue, municipal waste, kitchen waste, animal excreta, aquatic biomass (microalgae and macroalgae) are considered as the potential sources for ensuring the future energy supply [1,2]. Algal biomass is the promising source for biofuel production and conversion of these biomasses into biofuel such as biodiesel, bioethanol, biogas etc., have many benefits like less energy input steps; less production of sludge; cost-effective operation; and recycling of nutrients [3]. Algal biomass is known as third-generation biomass and classified as macroalgae and microalgae depending upon their size, availability and biochemical composition. Macroalgae are often observed in coastal marine habitats and about 10,000 different species have been analyzed till the date. There are more than 20,0000 species of microalgae and out of these only 30,000 have been analyzed in different studies. Algal biomass exhibit several qualities such as high growth rate, high capacity of CO_2 absorption, less or no requirement of arable land and farming facilities [4,5]. Moreover, algal biomass is considered as the efficient carbon-capturing biomass; its average photosynthesis efficiency is about 6-8%, while it is only 1.8-2.2% of the total terrestrial biomass [6]. Microalgae also utilise the inorganic substrate such as nitrogen, phosphorus, iron etc., as nutrients and this property make them a potential agent to absorb these impurities from waste water. Besides, some microalgal species can grow in concentrated salt water, and these halophilic microalgae do not compete for freshwater. Some of the marine microalgal species are being utilised to treat natural seawater contaminated with nitrogen and phosphorus, supplied by commercial fertilizers. In this way, these microalgal species provide dual benefits; feedstock for biofuel production and remediation for wastewater [7,8]. It was reported that 100 tonnes of algae absorbs almost 183 tonnes CO_2, hence utilisation of algal biomass to produce biofuel has hidden advantage of reducing the CO_2 from the environment, emitted by the use of fossil fuel [9]. Moreover, algal biomass is recommended for reducing the toxic substances from various elements of our environment and producing renewable biofuel. The growth and biochemical composition of algal biomass are sensitive to various factors such as light intensity, temperature, salt content, and nutrients availability [5,10].

2. CLASSIFICATION OF ALGAL BIOMASS

2.1 Characteristics of Macroalgae

Macroalgae are found in coastal marine habitats and till the date, approximately 10,000 varieties are documented with sizes ranges from millimetres to meters. The dry matter in macroalgae ranges between 10-15% of wet biomass, and almost 60% of the dry matters withthe carbohydrate content of almost 60% of dry mass [6]. Macroalgae are categorised into three categories depending upon the variation in photosynthetic pigmentation; brown algae (Phaeophyta), green algae (Chlorophyta) and red algae (Rhodophyta). Red and brown algae are generally found in warm water and temperate to cold or very cold water bodies, respectively, while green algae can grow in any kind of water temperature bodies [6,11,12]. Cell wall composition also varies according to the type of macroalgae. Green macroalgae comprise ulvan and xylan, while red algae contain alginate and fucoidan, as the main constituents. Some macroalgae also contain starch and cellulose such as chlorophycean, mannitol, floridean, and laminarin in their cell wall envelope [10,13]. Macroalgae contain easily hydrolyzable proteins and sugars, high hemicellulose and very low lignin content, these properties make them a potential feedstock for biofuel production.

2.2 Characteristics of Microalgae

Microalgae are a group of photosynthetic eukaryotes (Chlorophyta, red algae (Rhodophyta), diatoms (Bacillariophyta)) and prokaryotes (cyanobacteria). Microalgal cell envelope contains polysaccharide, glycolipids and glycoproteins as well as biopolymer such as pectin, cellulose, and hemicelluloses [14,15]. Cell wall constituents such as proline, hydroxyproline, albanians, uronic acid, sporopollenin, carotenoids, and glucosamine add to the recalcitrance of the microalgal biomass [16,17]. Microalgae are emerging as potential microorganism for detoxifying the variety of wastewater effluents loaded with toxic heavy metals and other undesired elements [18,19]. Many researchers studied the decontaminating properties of different microalgae on the industrial and municipality effluents. The growth rate of microalgae is very high (doubling time <24h), however, its productivity is very sensitive to the nutrients available in wastewater [20].

3. CULTIVATION OF ALGAL BIOMASS

3.1 Cultivation of Microalgae

Cultivation of microalgaeis performed in three metabolic ways such as phototrophic, heterotrophic, and mixotrophic. Phototrophic cultivation method utilizes the atmospheric CO_2 as carbon source and sunlight as the energy source, while heterotrophic cultivation system requires only carbon substrates such as glucose, glycerol and acetate [21]. Moreover, the heterotrophic system offers the advantage of higher microalgal biomass yield than the phototrophic system. Mixotrophic cultivation technique, facilitate the cultivation of microalgae under both phototrophic and heterotrophic conditions. Also, the mixotrophic technique provides improved growth rate and productivity compared to heterotrophic and phototrophic technique. Although heterotrophic technique comprises several benefits over phototrophic technique, it is less popular than phototrophic technique due to higher energy consumption, higher cost, and risk of contamination by other microorganisms. Being a less energy-intensive, and economic method of cultivation, phototrophic and autotrophic systems are mostly utilised for microalgae-based biofuel production [14,21].

A phototrophic microalga is generally cultivated in open raceway ponds (RWP) and photo bioreactors (PBR). RWP (Fig. 7.1) is the open shallow pond or channel-type structure (made up of concrete or plastics) with a water depth of 15-20 cm. Paddle wheel is equipped for mixing the culture medium that is further controlled by the baffles located at the flow channels.RWP technology has been utilised for mass cultivation ofmicroalgae since 1950s. The open environment operation in RWP systems suffers water loss due to evaporation and contamination by other microorganisms. Although RWP systems are economical and easy in operation, their usage is limited because of low biomass productivity compared to PBR systems [21,22].

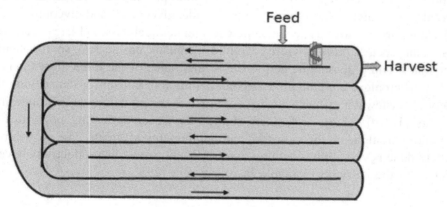

Figure 7.1: Surface View of Raceway Pond Microalgae Cultivation [21,22]

PBR technology facilitates the cultivation of desired microalgae species with minimum contamination, in a controlled environment with higher biomass productivity than the RWP technology. These closed bioreactors may be tubular, plate type or bubbled column. A plate PBR contains vertical or inclined rectangular boxes that are further divided into two parts to accelerate the flow of fluid inside the boxes. On the other hand, bubbled column PBR contains a vertical cylindrical column of transparent material. Among all the PBR systems, tubular PBR is considered as the most efficient PBR design, as it utilizes the maximum CO_2 and sunlight using minimum working area. A tubular system comprises a series of straight transparent solar energy collector tubes made up of transparent material (glass or plastic). The world's largest tubular PBR (Fig. 7.2) was installed by Bisan Tech., in Germany, which yields annually 100 tonnes of microalgae with the use of 20 independent modules with a total volume of approximately 700 m^3 [22, 23,24].

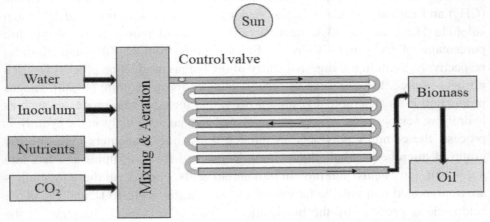

Figure 7.2: Tubular Photo Bioreactor for Microalgae Cultivation [22,23]

3.2 Cultivation of Macroalgae

Macroalgae (seaweeds) can directly be harvested in the open sea, without using controlled and designed facilities that are needed for microalgae cultivation [25]. However, macroalgae cultivation depends upon various important parameters such as nutrient supply, light source, temperature, etc. The distance of the harvesting sight from the suitable harbour (or distance from shore) also affects the economy of macroalgae cultivation as the transportation activity consumes energy and time. Macroalgae are generally cultivated in sheltered or lagoons to facilitate the easy and direct supply of nutrients of seawater. Macroalgae are commonly cultivated and harvested using three methods; Fixed off-bottom long line culture, Floating Rafts and Rock-based farming method. The fixed off-bottom long line culture method is generally utilised in near shore and relatively less deep waters with the substrate floating near the sea surface, Floating Rafts method is utilised in deeper waters with the substrate placed near the sea floor surface while Rock-based method is used at low tide region [26,27]. Currently, almost 42 countries cultivate macroalgae for different commercial purposes. China is the top in the list of macroalgae cultivating countries followed by Philippines, North Korea, Japan, and South Korea [27].

4. BIOGAS PRODUCTION FROM ALGAL BIOMASS

4.1 Anaerobic Digestion Process

Biogas is a product of complex biochemical degradation of organic substrate (carbohydrate protein, lipids, fat etc.) in anaerobic condition. It contains methane (CH_4) and carbon dioxide (CO_2) as major components with traces of hydrogen sulphide (H_2S), oxygen (O_2), water vapours (H_2O), and hydrocarbons. Volumetric percentage of CH_4 and CO_2 in the biogas varies from 55-70% and 30-45%, respectively. Semi liquid digested slurry of the biogas production process is a rich nitrogen source that can serve as organic manure for agriculture [14,28]. Biogas production is an anaerobic digestion process that includes four processes such as hydrolysis, acidogenesis, acetogenesis and methanogenesis. Firstly, in the hydrolysis process, the complex polymers are broken down into monomers by a specific group of microbes; carbohydrates into glucose and fructose, lipids and fats into long-chain fatty acids, and protein into amino acids. Further, all these monomers are transformed into volatile fatty acids (VFAs), acetic acids, CO_2, and H_2, in the acidogenesis process, by the biochemical action of acidogenic bacteria. In the third step (acetogenesis), acetogenic bacteria convert VFAs and H_2 into acetic acids. Subsequently, in the last step (methanogenesis), methanogenic bacteria utilise acetic acids to produce methane [1,29]. Figure 7.3 represents all the steps of anaerobic digestion process foralgal biomass.

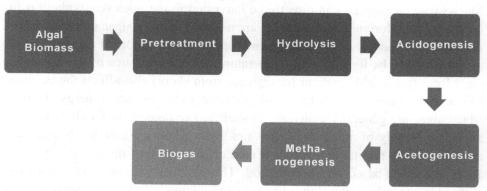

Figure 7.3: Anaerobic Digestion Process of Algal Biomass (Microalgae and Macroalgae)

4.2 Factor Affecting Anaerobic Digestion Process

The efficiency of anaerobic digestion process depends upon various factors such as nature of biomass (lignocellulosic biomass, kitchen waste, algae etc.), temperature, pH, organic loading rate (OLR), biomass loading, alkalinity, and hydraulic retention time (HRT).The optimum temperatures for thermophilic and mesophilic AD are considered to be 55 °C and 35 °C. The pH range of 6.6-7.4 is recommended for the efficient activity of all microbes in the AD process.OLR is also a crucial parameter which quantifies the amount of volatile solid (VS) needed per unit volume of anaerobic digester per unit time. OLR of 0.5-1.6 kgVS/m^3/day and 1.6-4.8 kgVS/m^3/day are considered to be the optimum OLR for slow rate AD and high rate AD, respectively.HRT is represented as the average time required for anaerobic bacteria to degrade or synthesis the fixed volume of biomass. Generally, HRT of 30-50 days is considered as the optimum HRT for pilot scale anaerobic digester while for lab-scale experiments it can be 15-30 days.

The ratio of carbon and nitrogen (C/N ratio) significantly affects the AD process. The C/N ratio of 20-30, is considered to be the optimum C/N ratio for the efficient AD. C/N ratio below the optimum value results in the production of higher ammonia (NH_3) which resists the growth of microbes while the C/N ratio above the optimum value results in the VFAs production, which imbalance the digester pH and eventually slows down the biogas production. Alkalinity is an important factor to provide the essential buffering capacity to countervail the accumulation of VFAs and to balance the optimum pH for efficient AD [10, 31, 31,32]. Equation 1, 2 and 3 represent the AD of carbohydrates, proteins, and lipid.

$C_6H_{10}O_5$ (Carbohydrates) + H_2O → $3CH_4$ + 3 CO_2 ... (1)

$C_{16}H_{24}O_5N_5$ (Proteins) + 14.5H_2O → 8.25CH_4 + 3.75 CO_2 + 4NH_3 ...(2)

$C_{50}H_{90}O_6$ (Lipids) + 24.5H_2O → 34.75CH_4 + 12.25 CO_2 ... (3)

4.3 Biogas Production from Microalgae and Macroalgae

Algal biomass has been utilized as a feed material for biogas production since the 1950s. Biomethane yield of green and brown microalgae was reported to be 227 L/kg VS and 262 L/kg VS, which is much higher than the lignocellulosic biomass (172 L/kg VS) and sugar crops (189 L/kg VS) [33]. Many researchers have evaluated the biomethane potential of algal biomass. Yen et al. [34] studied the anaerobic co-digestion (AcoD) of algal sludge (substrate with low C/N ratio) and waste paper (substrate with high C/N ratio) for balancing C/N ratio to improve biomethane yield of algal sludge. For the HRT of 10 days, AD of algal biomass resulted in methane production of 1.17L/L/day, which was 104.2% higher than the biomethane yield of algal sludge alone. The highest methane production of 1.607 L/L/d was noted for the OLR of 5 g VS/L/d and 60% VS of waste paper with algal sludge [34]. Jard et al. [13] evaluated the effect of thermo chemical pretreatment on biomethane yield of red macroalga Palmaria palmata (P.palmata). Thermochemical pretreatment was performed using NaOH and HCl at different temperature (20 °C -200 °C). P. palmata pretreated with 0.04 g NaOH/ gTS at 20°C, resulted the maximum methane yield (365 mL/g VS), which was 18.50% higher than the untreated one [13]. Prajapati et al. [35] studied the biomass and biogas production potential of three Chlorella species namely *C. minutissima, C. vulgaris and C. pyrenoidosa.* The highest biomass productivity of 0.98 g/L, was noticed for *C. pyrenoidosa* followed by *C. vulgaris* (0.92 g/L) and *C. minutissima* (0.90 g/L). The maximum biogas potential of 464L/kg VS (57.05% CH_4 content) was noted for *C. pyrenoidosa* followed by *C. vulgaris* (369 L/kg VS) and *C. minutissima* (340 L/kg VS) [35]. In another study, Prajapati et al. [36] evaluated the effect of fungal crude enzyme-based pretreatment on biomethane production of algal biomass. Algal biomass pretreated at optimised pretreatment condition resulted in improved methane production of 324.38 mL/g VS, which was 27.34% higher than the untreated algal biomass. Co-digestion of pretreated algal biomass(low C/N ratio) with cattle dung (high C/N ratio), increased the methane yield up to 413.89 mL/g VS, which was 62.48% and 27.59% higher than the mono-digestion of untreated and pretreated algal biomass, respectively [36]. Herrmann et al. [37] investigated the AcoD of microalgae *Arthrospira platensis* (*A. platensis*) and carbon-rich-biomass such as barley straw, beet silage and brown seaweed at the balanced C/N ratio of 25. No significant improvement in methane production was found with batch co-digestion experiment. However, AcoD in continuous digestion showed a significant synergistic effect on biomethane potential. Co-digestion of *A. platensis* and seaweed at OLR of 4 g VS/L/d at C/N ratio of 25 was found to be optimised condition for improved biomethane production [37].

There are many studies regardingbiogas potential of various species of macroalgae.Bruhn et al. [38] reported that biomethane yield of fresh and macerated *Ulva. Lactuca* (*U. Lactuca*) was 271 L/ kg VS. Also, dried *U. lactuca* yielded 5-9 times increase in weight-specific biomethane production compared to wet *U. lactuca.* Also,Allen et al.[39]evaluated the biomethane potential (BMP) of fresh *U. lactuca* (green seaweed) and dried washed, and macerated *U. lactuca.* Fresh *Ulva* was found to have BMP of 183 L/ kgVS while macerated *Ulva* showed the BMP of 250 L/ kgVS [39]. Oliveira et al. [40] applied the design of experiment technique (DOE) to find the strategies to improve the biomethane yield of macroalgae *Grailiaria vermiculophylla* (*G. vermiculophylla*). Physical pretreatment using washing and maceration improved the methane yield up to 481

L/ kg VS, while the thermo chemical pretreatment did not show any significant improvement in methane yield. Co-digestion of *G. vermiculophylla* and secondary sludge at the mixing ratio of 15:85 (TS/TS) improved the methane yield by 25% corresponding to methane yield of 605 L/kg VS [40]. Tedesco et al. [41] evaluated the performance of Hollander beater mechanical pretreatment on biomethane yield of *Laminariaceae* macroalgae. Pretreatment at optimised conditions was found to increase the biomethane yield up to 430 L/kg TS which was 53% higher than the methane yield of untreated macroalgae [41].

Kim et al. [42] reported the biomethane yield of *U. lactuca* as 430 L/ kg COD removed, which was very close to the estimated theoretical potential of the *U. Lactuca* [42]. Vergara-Fernández et al. [43]performed the two-phase AD of the macroalgae *Macrocystis pyrifera* and *Durvillea antarctica*, and their blend at mixing ratio of 1:1 (w/w). The two-phase anaerobic digester system was made up of an anaerobic sequencing batch reactor (ASBR) and an up-flow anaerobic filter (UAF). Both algae showed a similar biogas yield of 180 L/ kg dry algae/ day, with methane content of 65%. Gurung et al. [44] evaluated the biomethane potential of four seaweed, green algae, brown algae and fish viscera. After 60 days of digestion period, the cumulative biomethane yields of green and brown algae were observed to be 256 L/kg VS and 179 L/kg VS, respectively. Seaweed and fish viscera showed relatively lower methane yield of 102L/kg VS and 127 L/kg VS, respectively [44]. Jard et al. [45] performed AD experiment on microalgae *Palmaria palmate* (*P. palmate*) collected from French Brittany coasts (France). The results showed that the biochemical methane potential of *P. palmate* was 279 L/ kg VS [45].

5. BIOETHANOL PRODUCTION FROM ALGAL BIOMASS

Bioethanol is considered as an alternative of gasoline. According to the feedstock utilised, bioethanol is categorised as first, second and third-generation bioethanol. Food crops such as wheat, barley, maize, sugarcane, molasses, sweet sorghum, potato etc., were extensively utilised for producing first-generation bioethanol, however, due to food security reason and high production cost, the focus was shifted towards the second generation bioethanol [46,47]. Lignocellulosic biomasses such as crop residues were started to utilise instead of food crops in the second generation bioethanol, however, due to low yield and high cost of bioethanol production, the third generation of bioethanol production came into existence.Algal biomass is one of the main sources of the third generation of biomass. Besides, it does not compete with food crops as it doesn't require arable land to grow. It can be cultivated in the harsh environmental condition such as saline water, wastewater of industries, it absorbs 500-700% more CO_2 than the lignocellulosic or woody biomass [47,48,49]. Algal biomass yields two and five times higher bioethanol than the sugarcane and corn, respectively. In the bioethanol fermentation process, bacteria or yeast utilises carbohydrate part of algal biomass for bioethanol production. Water and CO_2 are produced as secondary products in bioethanol fermentation [50]. The Eq. (4) represents the conversion of glucose into bioethanol:

$$C_6H_{12}O_6 \rightarrow CH_3CH_2OH + CO_2 \quad \ldots (4)$$

5.1 Bioethanol Fermentation Process

Bioethanol production from algal biomass comprises several steps namely, pretreatment, hydrolysis, fermentation and distillation. Among all these steps pretreatment is the important step to break the rigid cell wall of algae and extract the carbohydrate from algae cell wall[51],[52]. Physical, enzymatic and chemical pretreatment is the most common pretreatment methods. Steam explosion and ultrasonic pretreatment were extensively utilised physical pretreatments. However, these physical pretreatment techniques are energy and cost-intensive. In contrast, chemical pretreatment using various acid and alkali chemicals such as H_2SO_4, HCl, NaOH, NH_4OH, $CaOH_2$ are found to be effective, easier and economical for pretreating algal biomass. However, some acidic conditions during the chemical pretreatment may initiate the decomposition of sugar forming undesirable compounds that further impede fermentation process. In contrast, enzymatic pretreatment can produce higher glucose yields in an environmentally friendly way, without forming inhibitory compounds [53,54,55,56]. After the first step of pretreatment the bioethanol production can be done in two ways; by separate hydrolysis and fermentation (SHF) (Fig.7.4), and simultaneous saccharification and fermentation (SSF) (Fig.7.5). In SHF, algae are first hydrolysed and the hydrolysed biomass then subjected to fermentation process utilising microorganisms such as bacteria or yeast. In contrast, in the SSF process hydrolysis and fermentation is performed simultaneously in a single step. SSF process provides many benefits such as reduction of the product inhibitors, process time, capital cost and simplified the reaction procedure. After the fermentation process, bioethanol is separated from water by fractional distillation technique. In this technique, the unfiltered bioethanol is boiled, and due to the lower boiling point of 78.3 °C, liquid bioethanol is converted into bioethanol steam before water, this bioethanol steam is captured with the purity of 95% and water is separated using condensation process. In the large scale industries and biorefineries, this distillation is performed using continuous distillation column system [27,53,57].

Algal Biomass: A Promising Source for Future Bioenergy Production

Figure 7.4: Separate Hydrolysis and Fermentation Process for Bioethanol Production

Figure 7.5: Simultaneous Saccharification and Fermentation Process for Bioethanol Production

5.2 Bioethanol Potential of Microalgae and Macroalgae

Harun et al. [58] investigated the effect of alkaline pretreatment on bioethanol potential of macroalgae *Chlorococcum infusionum*, using NaOH as a chemical reagent. The pretreatment process was analysed for three parameters such as NaOH concentration, pretreatment time and temperature. The maximum bioethanol production of 0.26 g/ g algae was obtained for the algae pretreated with 0.75% (w/v) NaOH, at 120 °C for 30 min. of pretreatment time [58]. Ho et al. [56] evaluated the bioethanol potential of carbohydrate-rich microalga *Chlorella vulgaris* FSP-E strain (*C.vulgaris*). The SHF and SSF techniques of bioethanol production converted the enzymatic *C. vulgaris* algae hydrolysate into bioethanol with a 79.9 % and 92.3% theoretical yield. Hydrolysis with 1% H_2SO_4 was observed to be effective for saccharifying the algae biomass as acidic hydrolysate of algae via SHF process, produced the bioethanol at a concentration of 11.7 g/L with a theoretical yield of 87.6% [56]. Wu et al. [59] performed sequential acid and enzyme hydrolysis of *Gracilaria* sp. red seaweed to enhance its fermentability and hence bioethanol production. Saccharification at optimal condition resulted in bioethanol concentration of 4.72 g/L with bioethanol productivity of 4.93 g/L/day. The finding of this study, confirmed the utility of *Gracilaria* sp. red seaweed for ethanol production on a large scale [59]. Tan et al. [60] performed SHF and SSF of seaweed solid waste (*Saccharomyces cerevisiae*) for bioethanol production via enzymatic hydrolysis. Bioethanol yield of 55.9% was obtained for the SHF process while it was 90.9% for SSF process [60]. Trivediet al. [61] evaluated the effect of enzymatic hydrolysis of green seaweed Ulva for improved bioethanol production. The fermentation of hydrolysate obtained from enzymatic hydrolysis performed with 2% (v/v) enzyme, resulted in bioethanol yield of 0.45 g/g reducing sugar with the conversion efficiency of 88.2% [61]. Meinita et al. [62] reported that bioethanol yield of acid hydrolysate red seaweed *Carrageenophyte Kappaphycus alvarezii* powder as 0.21 g/g galactose with a theoretical yield of 41% [62].

6. BIODIESEL PRODUCTION FROM ALGAL BIOMASS

Biodiesel is the liquid biofuel that comprises methyl esters of long-chain fatty acids. It is generally derived from transesterification of vegetable oils, waste cooking oils, animal fats or lipids of algal biomass using alcohol and alkali and acids as a catalyst [22,63]. Utilization of biodiesel in the place petroleum diesel comprises many advantages; it is ready to use fuel in an existing diesel engine with slight modification in engines, it can be utilised with or without blending with petroleum diesel. As it is derived from biomass, it does not add to atmospheric

CO_2. Moreover, the use of biodiesel instead of petroleum diesel reduces sulphur, shoot and unburned hydrocarbons. Biodiesel has twice the viscosity of petroleum diesel, this property of biodiesel improves engine life, when it is utilised in the engine as a fuel. Moreover, biodiesel could significantly improve the rural economy [22,64,65,66].

6.1 Biodiesel Production Process

Biodiesel production from algal biomass starts from collection of wet algal biomass, followed by dewatering, drying, pretreatment, lipid extraction, fractionation, and transesterification process. In the dewatering process, biomass paste is obtained by removing the water from the depleted cell culture. Forward osmosis was reported as the less energy consuming membrane separation method for dewatering process. Algal biomass paste or slurry obtained from dewatering process is further dried using various drying process such as solar drying, spray drying, freeze-drying and spray drying [67,68,69].

6.1.1. Pretreatment

Pretreatment is an essential step in biodiesel production as it significantly affects the biodiesel productivity. Pretreatment of algal biomass is generally performed to disrupt the algae cell using physical (mechanical and thermal), chemical and biological or combined pretreatment technique. The pretreatment method is selected according to chemical characteristics of algal species [69,70]. Lipid extraction is the crucial step in the biodiesel production from algal biomass. Lipid extraction is generally carried out with two solvent extraction method such as Folch method and Bligh and Dyer method. In the Folch lipid extraction method, the chloroform-methanol mixture is utilised in a volume ratio of 2:1, while in Bligh and Dyer method the same mixture is utilised in a volume ratio of 1:2 [71,72]. The Bligh and Dyer method was reported to be utilised in commercial-scale lipid extraction process from algae [73].

6.1.2 Fractionation

Fractionation is performed to separate the desired material from a mixture having impurity of undesired material, based on product size, solubility, volatility and ionic strength. The extracted lipid may contain the impurities of proteins, carotenes, non-acylglycerols, neutral lipids and chlorophylls. Molecular weight distribution is one of the most utilised parameters to separate the desired product by a different immiscible layer of liquids. Membrane separation can also be an efficient method, in which separation is conducted by changing the membrane pore size and permeate flux rate, which subsequently facilitates the fractionation of desired product based on molecular size [67,74,75].

6.1.3 Transesterification

In the last, transesterification step, triglycerides of algae oil react with alcohol (methanol or ethanol) and produce fatty acid methyl ester or fatty acid ethyl ester with glycerol as products.The reaction is performed at atmospheric pressure and temperature of 60-70°C. Acid or alkali chemicals, such as NaOH or H_2SO_4, are often used as a catalyst to speed up the reaction rate and reduce the reaction time. After the transesterification process, the pure biodiesel is separated and impurities of free fatty acids, acid or alkaline catalyst and alcohol are removed in the purification process. It was reported that saponification during transesterification process hinders the reaction; however, it can be avoided by using supercritical transesterification method as it does not require a catalyst to boost up the reaction process [67,76]. Fig. 7.6 represents the process of biodiesel production from algal biomass.

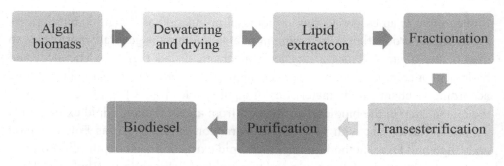

Figure 7.6: Biodiesel Production Process from Algal Biomass

Macroalgae is much suitable for bioethanol and biogas production than biodiesel production as most of the macroalgae species do not contain triglycerides. There are very few researches about biodiesel production from macroalgae and all these researches reported the much lower yields than the biodiesel yields of microalgae [77,78].

7. CONCLUSIONS

Algal biomass could be a promising source of biofuel production, however, there are still several challenges in various stages, from cultivation to final product. Cultivation, harvesting and dewatering are the energy and cost-intensive steps. To make all these processes energy-efficient and cost-effective, there is a need for further research considering various objectives such as the selection of effective algal strain, optimization of cultivation, harvesting and dewatering process. Although there are many lab-scale studies on conversion the algal biomass into biofuel via biochemical routes, there is a requirement of techno-economic studies to make all these conversion technologies feasible for industrial application.

REFERENCES

1. Kumar S, Paritosh K, Pareek N, Chawade A, Vivekanand V. De-construction of major Indian cereal crop residues through chemical pretreatment for improved biogas production/ : An overview. Renew Sustain Energy Rev. 2018; 90: 160–70. https://doi.org/10.1016/j.rser.2018.03.049.
2. Gandhi P, Kumar S, Paritosh K, Pareek N, Vivekanand V. Hotel generated food waste and its biogas potential/ : A case study of Jaipur city, India. Waste and Biomass Valorization. 2019; 10: 1459–68. https://doi.org/10.1007/s12649-017-0153-1.
3. Grosser A. The influence of decreased hydraulic retention time on the performance and stability of co-digestion of sewage sludge with grease trap sludge and organic fraction of municipal waste. J Environ Manage. 2017; 203: 1143–57. https://doi.org/10.1016/j.jenvman. 2017.04.085.
4. Dar RA, Arora M, Phutela UG. Optimization of cultural factors of newly isolated microalga Spirulina subsalsa and its co-digestion with paddy straw for enhanced biogas production. Bioresour Technol Reports. 2019; 5: 185–98. https://doi.org/ 10.1016/j.biteb.2019.01.009.
5. Saratale RG, Kuppam C, Mudhoo A, Saratale GD, Periyasamy S, Zhen G. Bioelectrochemical systems using microalgae – A concise research update. Chemosphere. 2017; 177: 35–43. https://doi.org/10.1016/j. Chemosphere. 2017.02.132.
6. Chen H, Zhou D, Luo G, Zhang S, Chen J. Macroalgae for biofuels production: Progress and perspectives. Renew Sustain Energy Rev. 2015; 47: 427–37. https://doi.org/10.1016/j.rser.2015.03.086.
7. Nguyen DD, Ngo HH, Yoon YS. Effect of internal recycling ratios on biomass parameters and simultaneous reduction of nitrogen and organic matter in a hybrid treatment system. Ecol Eng. 2016. https://doi.org/10.1016/j.ecoleng. 2015.11.054.

8. Gupta SK, Ansari FA, Shriwastav A, Sahoo NK, Rawat I, Bux F. Dual role of Chlorella sorokiniana and Scenedesmus obliquus for comprehensive wastewater treatment and biomass production for bio-fuels. J Clean Prod. 2016. https://doi.org/10.1016/j.jclepro.2015.12.040.

9. Doucha J. Utilization of flue gas for cultivation of microalgae (Chlorella sp .) in an outdoor open thin-layer photobioreactor Utilization of flue gas for cultivation of microalgae (Chlorella sp .) in an outdoor open thin-layer photobioreactor. 2014. https://doi.org/10.1007/s10811-005-8701-7.

10. Oliveira J V., Alves MM, Costa JC. Design of experiments to assess pre-treatment and co-digestion strategies that optimize biogas production from macroalgae Gracilaria vermiculophylla. Bioresour Technol. 2014; 162: 323–30. https:// doi.org/ 10.1016/j.biortech.2014.03.155.

11. Ganesh Saratale R, Kumar G, Banu R, Xia A, Periyasamy S, Dattatraya Saratale G. A critical review on anaerobic digestion of microalgae and macroalgae and co-digestion of biomass for enhanced methane generation. Bioresour Technol. 2018; 262: 319–32. https://doi.org/10.1016/j.biortech.2018.03.030.

12. Shobana S, Saratale GD, Pugazhendhi A, Arvindnarayan S, Periyasamy S, Kumar G. Fermentative hydrogen production from mixed and pure microalgae biomass: Key challenges and possible opportunities. Int J Hydrogen Energy. 2017; 42: 26440–53. https://doi.org/10.1016/j.ijhydene.2017.07.050.

13. Jard G, Dumas C, Delgenes JP, Marfaing H, Sialve B, Steyer JP. Effect of thermochemical pretreatment on the solubilization and anaerobic biodegradability of the red macroalga Palmaria palmata. Biochem Eng J. 2013; 79: 253–8. https:// doi.org / 10.1016/j.bej.2013.08.011.

14. Zabed HM, Akter S, Yun J, Zhang G, Zhang Y, Qi X. Biogas from microalgae: Technologies, challenges and opportunities. Renew Sustain Energy Rev. 2020; 117: 109503. https://doi.org/10.1016/j.rser.2019.109503.

15. González-Fernández C, Sialve B, Bernet N, Steyer JP. Comparison of ultrasound and thermal pretreatment of Scenedesmus biomass on methane production. Bioresour Technol. 2012; 110: 610–6. https://doi.org/10.1016/j.biortech.2012.01.043.

16. Ward AJ, Lewis DM, Green FB. Anaerobic digestion of algae biomass: A review. Algal Res. 2014; 5: 204–14. https://doi.org/10.1016/j.algal.2014.02.001.

17. Kwietniewska E, Tys J. Process characteristics, inhibition factors and methane yields of anaerobic digestion process, with particular focus on microalgal biomass fermentation. Renew Sustain Energy Rev. 2014; 34: 491–500. https:// / doi.org /10.1016 /j.rser.2014.03.041.

18. Zhou W, Min M, Li Y, Hu B, Ma X, Cheng Y. *et al.* A hetero-photoautotrophic two-stage cultivation process to improve wastewater nutrient removal and enhance algal lipid accumulation. Bioresour Technol. 2012; 110: 448–55. https:// doi.org / 10.1016 / j.biortech.2012.01.063.
19. Li Y, Chen YF, Chen P, Min M, Zhou W, Martinez B. Characterization of a microalga Chlorella sp. well adapted to highly concentrated municipal wastewater for nutrient removal and biodiesel production. Bioresour Technol. 2011; 102: 5138–44. https://doi.org/10.1016/j.biortech.2011.01.091.
20. Karunanithy C, Muthukumarappan K. Influence of extruder temperature and screw speed on pretreatment of corn stover while varying enzymes and their ratios. Appl Biochem Biotechnol. 2010; 162: 264–79. https://doi.org/10.1007/s12010-009-8757-y.
21. Moreno GL, Adjalle K, Barnabe S, Raghavan GSV. Microalgae biomass production for a biorefinery system: recent advances and the way towards sustainability. Renewable and Sustainable Energy Reviews. 2017; 76: 493–506.
22. Abomohra AE, Jin W, Tu R, Han S. Microalgal biomass production as a sustainable feedstock for biodiesel/ : Current status and perspectives. Renew Sustain Energy Rev. 2016; 64: 596–606. https://doi.org/10.1016/j.rser.2016.06.056.
23. Bajhaiya A, Mandotra S, Suseela M, Toppo K, Ranade S. Algal biodiesel The next generation biofuel for India. Asian J Exp Biol Sci. 2010.
24. Carvalho AP, Meireles LA, Malcata FX. Microalgal reactors: A review of enclosed system designs and performances. Biotechnol Prog. 2006. https://doi.org / 10.1021 / bp060065r.
25. Surendhiran D, Vijay M, Sivaprakash B, Sirajunnisa A. Kinetic modeling of microalgal growth and lipid synthesis for biodiesel production. Biotech. 2015. https://doi.org/10.1007/s13205-014-0264-3.
26. Krishnan M, Narayanakumar R. Social and economic dimensions of carrageenan seaweed farming. FAO Fish Aquac Tech Pap. 2013.
27. Razack A, Surendhiran D. Algae – a quintessential and positive resource of bioethanolproduction/ : A comprehensive review. Renew Sustain Energy Rev. 2016; 66: 248–67. https://doi.org/10.1016/j.rser.2016.07.024.
28. Braun R. Anaerobic digestion: A multi-faceted process for energy, environmental management and rural development. Improv. Crop Plants Ind. End Uses. 2007. https://doi.org/10.1007/978-1-4020-5486-0_13.

29. Deublein D, Steinhauser A. Biogas from waste and renewable resources: an introduction, Second Edition. 2010. https://doi.org/10.1002/9783527632794.

30. Brown D, Shi J, Li Y. Comparison of solid-state to liquid anaerobic digestion of lignocellulosic feedstocks for biogas production. Bioresour Technol. 2012. https://doi.org/10.1016/j.biortech.2012.08.051.

31. Callander IJ, Barford JP. Anaerobic digestion of high sulphate cane juice stillage in a tower fermenter. Biotechnol Lett. 1983. https://doi.org/10.1007/BF01386497.

32. Naik L, Gebreegziabher Z, Tumwesige V, Balana BB, Mwirigi J, Austin G. Factors determining the stability and productivity of small scale anaerobic digesters. Biomass and Bioenergy. 2014. https://doi.org/10.1016/j.biombioe.2014.01.055.

33. Song M, Pham HD, Seon J, Woo HC. Overview of anaerobic digestion process for biofuels production from marine macroalgae: a developmental perspective on brown algae. Korean J Chem Eng. 2015. https://doi.org/10.1007/s11814-015-0039-5.

34. Yen HW, Brune DE. Anaerobic co-digestion of algal sludge and waste paper to produce methane. Bioresour Technol. 2007; 98: 130–4. https:// doi.org/ 10.1016 / j.biortech.2005.11.010.

35. Prajapati SK, Malik A, Vijay VK. Comparative evaluation of biomass production and bioenergy generation potential of Chlorella spp. through anaerobic digestion. Appl Energy. 2014; 114: 790–7. https://doi.org/10.1016/j.apenergy.2013.08.021.

36. Prajapati SK, Malik A, Vijay VK, Sreekrishnan TR. Enhanced methane production from algal biomass through short duration enzymatic pretreatment and codigestion with carbon rich waste. RSC Adv 2015; 5: 67175–83. https: // doi.org / 10.1039 / c5ra12670c.

37. Herrmann C, Kalita N, Wall D, Xia A, Murphy JD. Optimised biogas production from microalgae through co-digestion with carbon-rich co-substrates. Bioresour Technol. 2016; 214: 328–37. https://doi.org/10.1016/j.biortech.2016.04.119.

38. Bruhn A, Dahl J, Nielsen HB, Nikolaisen L, Rasmussen MB, Markager S. et al. Bioenergy potential of Ulva lactuca: Biomass yield, methane production and combustion. Bioresour Technol. 2011. https://doi.org/10.1016/j.biortech.2010.10.010.

39. Allen E, Browne J, Hynes S, Murphy JD. The potential of algae blooms to produce renewable gaseous fuel. Waste Manag. 2013. https://doi.org /10.1016 / j.wasman. 2013.06.017.

40. Oliveira JV, Alves MM, Costa JC. Bioresource technology design of experiments to assess pre-treatment and co-digestion strategies that optimize biogas production from macroalgae Gracilaria vermiculophylla. Bioresour Technol. 2014; 162: 323–30. https://doi.org/10.1016/j.biortech.2014.03.155.

41. Tedesco S, Marrero Barroso T, Olabi AG. Optimization of mechanical pre-treatment of Laminariaceae spp. biomass-derived biogas. Renew Energy. 2014. https://doi.org / 10.1016/j.renene.2013.08.023.

42. Kim J, Jung H, Lee C. Shifts in bacterial and archaeal community structures during the batch biomethanation of Ulva biomass under mesophilic conditions. Bioresour Technol. 2014. https://doi.org/10.1016/j.biortech.2014.07.041.

43. Ergara-Fernández A, Vargas G, Alarcón N, Velasco A. Evaluation of marine algae as a source of biogas in a two-stage anaerobic reactor system. Biomass and Bioenergy. 2008. https://doi.org/10.1016/j.biombioe.2007.10.005.

44. Gurung A, Van Ginkel SW, Kang WC, Qambrani NA, Oh SE. Evaluation of marine biomass as a source of methane in batch tests: A lab-scale study. Energy. 2012. https://doi.org/10.1016/j.energy.2012.04.005.

45. Jard G, Marfaing H, Carrère H, Delgenes JP, Steyer JP, Dumas C. French Brittany macroalgae screening: Composition and methane potential for potential alternative sources of energy and products. Bioresour Technol. 2013. https://doi.org/ 10.1016/ j.biortech.2013.06.114.

46. Singh A, Nigam PS, Murphy JD. Renewable fuels from algae: an answer to debatable land based fuels. Bioresour Technol. 2011. https://doi.org/ 10.1016 / j.biortech.2010.06.032.

47. John RP, Anisha GS, Nampoothiri KM, Pandey A. Micro and macroalgal biomass: a renewable source for bioethanol. Bioresour Technol. 2011. https://doi.org/ 10.1016/ j.biortech.2010.06.139.

48. Hong IK, Jeon H, Lee SB. Comparison of red, brown and green seaweeds on enzymatic saccharification process. J Ind Eng Chem. 2014. https://doi.org/ 10.1016 / j.jiec.2013.10.056.

49. Powers SE, Baliga R. Sustainable algae biodiesel production in cold climates. Int J Chem Eng. 2010. https://doi.org/10.1155/2010/102179.

50. Zhang W, Zhang J, Cui H. The isolation and performance studies of an alginate degrading and ethanol producing strain. Chem Biochem Eng Q. 2014. https://doi.org/ 10.15255/CABEQ.2013.1888.

51. Choi SP, Nguyen MT, Sim SJ. Enzymatic pretreatment of Chlamydomonas reinhardtii biomass for ethanol production. Bioresour Technol. 2010. https://doi.org/ 10.1016/j.biortech.2010.02.026.

52. Kang Q, Appels L, Tan T, Dewil R. Bioethanol from lignocellulosic biomass: Current findings determine research priorities. Sci World J. 2014. https:// doi.org/ 10.1155/ 2014/298153.

53. Kim HM, Wi SG, Jung S, Song Y, Bae HJ. Efficient approach for bioethanol production from red seaweed Gelidium amansii. Bioresour Technol. 2015. https:// doi.org/10.1016/j.biortech.2014.10.050.

54. Nahak S, Nahak G, Pradhan I, Sahu RK. Bioethanol from Marine Algae/ : A Solution to Global Warming Problem. J Appl Env Biol Sci. 2011.

55. Harun R, Danquah MK, Forde GM. Microalgal biomass as a fermentation feedstock for bioethanol production. J Chem Technol Biotechnol. 2010. https://doi.org/ 10.1002/jctb.2287.

56. Ho SH, Huang SW, Chen CY, Hasunuma T, Kondo A, Chang JS. Bioethanol production using carbohydrate-rich microalgae biomass as feedstock. Bioresour Technol. 2013. https://doi.org/10.1016/j.biortech.2012.10.015.

57. Tamayo JP, Rosario EJD. Chemical analysis and utilization of sargassum sp. As substrate for ethanol production. Iran Journal of Energy and Environment.2014; 5(2): 202-208.

58. Harun R, Jason WSY, Cherrington T, Danquah MK. Exploring alkaline pre-treatment of microalgal biomass for bioethanol production. Appl Energy. 2011. https://doi.org/ 10.1016/j.apenergy.2010.10.048.

59. Wu FC, Wu JY, Liao YJ, Wang MY, Shih IL. Sequential acid and enzymatic hydrolysis in situ and bioethanol production from Gracilaria biomass. Bioresour Technol. 2014. https://doi.org/10.1016/j.biortech.2014.01.024.

60. Tan IS, Lee KT. Enzymatic hydrolysis and fermentation of seaweed solid wastes for bioethanol production: An optimization study. Energy. 2014. https://doi.org/ 10.1016/j.energy.2014.04.080.

61. Trivedi N, Gupta V, Reddy CRK, Jha B. Enzymatic hydrolysis and production of bioethanol from common macrophytic green alga Ulva fasciata Delile. Bioresour Technol 2013. https://doi.org/10.1016/j.biortech.2013.09.103.

62. Meinita MDN, Kang JY, Jeong GT, Koo HM, Park SM, Hong YK. Bioethanol production from the acid hydrolysate of the carrageenophyte Kappaphycus alvarezii (cottonii). J Appl Phycol. 2012. https://doi.org/10.1007/s10811-011-9705-0.

63. Abomohra AEF, El-Sheekh M, Hanelt D. Pilot cultivation of the chlorophyte microalga Scenedesmus obliquus as a promising feedstock for biofuel. Biomass and Bioenergy. 2014. https://doi.org/10.1016/j.biombioe.2014.03.049.

64. Ramadhas AS, Jayaraj S, Muraleedharan C. Theoretical modeling and experimental studies on biodiesel-fueled engine. Renew Energy. 2006. https://doi.org/10.1016/j.renene.2005.09.011.

65. Mohamed Musthafa M. Synthetic lubrication oil influences on performance and emission characteristic of coated diesel engine fuelled by biodiesel blends. Appl Therm Eng. 2016. https://doi.org/10.1016/j.applthermaleng.2015.12.011.

66. Aresta M, Dibenedetto A, Carone M, Colonna T, Fragale C. Production of biodiesel from macroalgae by supercritical CO_2 extraction and thermochemical liquefaction. Environ Chem Lett. 2005. https://doi.org/10.1007/s10311-005-0020-3.

67. Yew GY, Lee SY, Show PL, Tao Y, Law CL, Nguyen TTC. Recent advances in algae biodiesel production: From upstream cultivation to downstream processing. Bioresour Technol Reports. 2019. https://doi.org/10.1016/j.biteb.2019.100227.

68. Hoover LA, Phillip WA, Tiraferri A, Yip NY, Elimelech M. Forward with osmosis: Emerging applications for greater sustainability. Environ Sci Technol 2011; 45: 9824–30. https://doi.org/10.1021/es202576h.

69. Passos F, Uggetti E, Carrère H, Ferrer I. Pretreatment of microalgae to improve biogas production: a review. Bioresour Technol. 2014; 172: 403–12. https://doi.org/10.1016/j.biortech.2014.08.114.

70. González-Fernández C, Sialve B, Bernet N, Steyer JP. Impact of microalgae characteristics on their conversion to biofuel. Part II: Focus on biomethane production. Biofuels, Bioprod Biorefining. 2012. https://doi.org/10.1002/bbb.337.

71. Bligh EG, Dyer WJ. A rapid method of total lipid extraction and purification. Can J Biochem Physiol. 1959. https://doi.org/10.1139/o59-099.

72. Folch J, Lees M, Sloane Stanley GH. A simple method for the isolation and purification of total lipides from animal tissues. J Biol Chem. 1957. https://doi.org/10.3989/scimar.2005.69n187.

73. Breil C, Abert Vian M, Zemb T, Kunz W, Chemat F. "Bligh and Dyer" and Folch methods for solid–liquid–liquid extraction of lipids from microorganisms. Comprehension of solvatation mechanisms and towards substitution with alternative solvents. Int J Mol Sci. 2017. https://doi.org/10.3390/ijms18040708.

74. Singh R. Membrane Technology and Engineering for Water Purification, Application, Systems Design and Operation. 2015.

75. Lee SY, Khoiroh I, Ling TC, Show PL. Enhanced recovery of lipase derived from Burkholderia cepacia from fermentation broth using recyclable ionic

liquid/polymer-based aqueous two-phase systems. Sep Purif Technol. 2017. https://doi.org /10.1016/ j.seppur.2017.01.047.

76. Zhang X, Yan S, Tyagi RD, Drogui P, Surampalli RY. Ultrasonication aided biodiesel production from one-step and two-step transesterification of sludge derived lipid. Energy. 2016. https://doi.org/10.1016/j.energy.2015.11.016.

77. Afify AEMMR, Shalaby EA, Shanab SMM. Enhancement of biodiesel production from different species of algae. Grasas Y Aceites 2010. https:// doi.org/ 10.3989/ gya.021610.

78. Maceiras R, Rodrí guez M, Cancela A, Urréjola S, Sánchez A. Macroalgae: raw material for biodiesel production. Appl Energy. 2011. https://doi.org/10.1016/ j.apenergy.2010.11.027.

CHAPTER - 8

MICRO ALGAE PRODUCTION FOR BIO FUEL GENERATION

Swapnaja K. Jadhav*, Anil K. Dubey, Mayuri Gupta, Sachin Gajendra and Panna Lal Singh

Agricultural Energy and Power Division
Central Institute of Agricultural Engineering, Navi Bagh,
Bhopal-462038, Madhya Pradesh, India
*Corresponding Author

Micro algae are a diverse group of photosynthetic unicellular organisms and are emerging as a promising source of renewable biomass for the production of bio fuels and valuable chemicals. The importance of micro algae and their products has significantly increased in the fields of human health, food, feed and bio fuels. Micro algae based bio fuel is considered as a third generation renewable fuel with plenty of advantages over the fossil fuels along with the CO_2 mitigation. Though micro algae can grow in their natural habitat, for production and harvesting biomass, it needs isolated structures. There are some established popular structures used worldwide as i) Open pond (Race way and circular pond), ii) closed systems called as photo bioreactors. Harvesting method can be selected depending on culture condition, size, shape of algae and the end product utilization. These techniques include filtration, centrifugation, coagulation, floatation, flocculation using natural; chemical; electro flocculants, and combination of methods. This chapter gives brief about introduction of micro algae production, Lipid producing micro algae, factors affecting its growth and production, design considerations for open pond systems and its application in bio fuel generation, etc.

1. INTRODUCTION

First generation bio fuel processes are useful but limited in most cases: there is a threshold above which they cannot produce enough biofuel without threatening food supplies and biodiversity. Ethanol/biodiesel from food-based crops competes for scarce cropland, fresh water, and fertilizers. Second generation biofuels are made from lignocelluloses biomass or woody crops, agricultural residues or waste, which makes it harder to extract the required fuel (ethanol, butanol). Third generation bio fuel are made using non-arable land, based on integrated technologies that produce a feedstock [1] as well as a fuel (or fuel precursor, such as pure vegetable oil), and require the destruction of biomass. Microalgae based bio fuels are gaining wide attention among bio fuel industries and researchers for its unexploited benefits. Micro algae based bio fuel is considered as a third generation renewable fuel with plenty of advantages over the fossil fuels along with the CO_2 mitigation.

Micro algae biomass is not only raw material for Biodiesel, Butanol, Gasoline, Methane, Ethanol, jet fuelbut also a potential source of other high value products. Several developed countries already have micro algae based bio fuel plants and waste water treatments plants. India has research and development based pilot plants for bio fuel production at leading research institutes and research is being focused on micro algal biomass production from municipal waste water treatment, marine water, dairy or food processing effluent and using low cost growth media in fresh water. With the results of it, microalgae soon will reserve its place among the leading renewable and carbon capture technologies in India.

2. WHAT ARE MICRO ALGAE?

Micro algae are a diverse group of photosynthetic unicellular organisms having size ranging between 3 to 30 μm and of different shapes. Microalgae include eukaryotic and prokaryotic cyanobacteria and capable of growing in very different environments. It has plenty of strains which can grow in any type of water and even on a moist surface having adequate nutrients, pH, aeration and light intensity. Like terrestrial plants, micro algae use light energy to convert carbon dioxide through photosynthesis into chemical energy in the form of biomass. Micro algae are reportedly having higher photosynthesis efficiency and have ability to utilize 1.6 to 2 kg of CO_2 per kg of algal biomass during process. The process of microalgae cultivation and its utilization for biofuel is shown in Fig. 8.1.

Compared to terrestrial oilseed crops, micro algae can produce 5 to 10 times biomass considering land, other resources and time requirement due to higher growth rate and short life cycle of algae. Unlike conventional crops, microalgae are capable to grow in poor quality, salty, brackish and waste water. Thus, algae being a non food source have a big potential of recycling liquid waste or gaseous waste streams back to organic biomass efficiently than conventional crops by utilizing nitrates, phosphates and other inorganic elements. These microalgae contribute as largest oxygen producers on the earth and can produce many high value compounds, proteins, fine chemicals, omega-3 fatty acids, anti-oxidants and pigments which can be used in production of pharmaceuticals, cosmetics, nutrition, food and feed [2, 3]. Besides this, oil producing algae contain lipids up to 60% of their dry mass making them potential future biofuel. It is necessary to have an infrastructure –downstream processes to economically produce large amount of lipid producing microalgae biomass to produce algal oils [4].

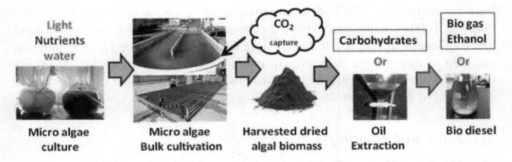

Figure 8.1: Microalgae Cultivation and its Utilization for Biofuel

Table 8.1: Oil Content of Different Micro-Algal Species

	Microalgae	Lipid Content (% dry wt)
Marine Water	Botryococcus braunii	25-75
	Crypthecodinium sp.	28-32
	Cylindrotheca sp.	20
	Daunaliella primolecta	16-37
	Isochrysis sp.	23
	Monallanthus salina	25-33
	Nitzschia sp.	>20
	Phaeodactylum tricornutum	20-35
	Schizochytrium sp.	31-68
	Tetraselmis suecica	35-54
Fresh Water	Nannochloris sp.	45-47
	Neochloris oleoabundans	50-77
	Chlorella vulgaris	14-56
Brackish Water	Nannochloropsis sp.	15-23
	Chlorella Sorokiniana	22-44

3. FACTORS INFLUENCING MICROALGAE GROWTH

Different physical, environmental and biotic factors can directly and indirectly affect the growth and biomass production of micro algae.

3.1 Physical Factor

3.1.1 Light Intensity

Since, Micro algae are a photosynthetic organism; it requires a certain amount of light for natural growth. Various microalgae can grow at light intensity in the range of 1000 to 8000 flux. Above a certain value, continue increasing in light intensity level will decrease the microalgae growth rate. Avoiding photo inhibition can help to increase the daily growth rate of microalgae biomass [4]. There are phenomena of heterotrophic and mixotrophic microalgae which can grow in the dark using a carbon source.

3.1.2 pH and Salinity

The pH value will also affect the growth rate of microalgae; it will be easier for microalgae to capture CO_2 in the atmosphere when the growing condition is alkaline (7, 8, 9), which can produce more biomass [5, 6]. Content of chlorophyll of microalgae will decrease when the pH value goes from 8.5 to 9.5 [7,8].

3.1.3 Nitrogen/Phosphorus Nutrient

Nitrogen is an essential element for growth of microalgae. The nitrogen and phosphorus ratio (N/P) directly affect the growth of microalgae, composition of cells and nutrient uptake. According to Redfield's law, the atomic ratio C: N: P in algae cells is 106:16:1, when the N/P ratio exceeds 16, concentration of phosphorus will be considered as a limited factor. Therefore, it is estimated that 88 kg N and 12 kg P are required for each metric ton (1000 kg) of dry algae biomass produced [9]. However, different species of microalgae has different atomic ratio in cells, the requirement for nitrogen and phosphorus will be various.

3.1.4 Temperature

With the light intensity changing, temperature is an environmental factor which indirectly affects growth of microalgae [10]. After temperature is higher than 11°C up to 25°C, the relationship between temperature and growth of microalgae is linear. Temperature determines the activity and reaction rates and affects the growth of microalgae and to limit its distribution.

3.1.5 Carbon Dioxide

This is also an important factor as carbon dioxide is dissolved in the media but is necessary for micro algae for photosynthesis. For this reason, many researchers believe that exhaust gases from power plants or industrial facilities will be used to provide CO_2 for large-scale algae production facilities. These exhaust gases can contain between 10 to 30 times higher CO_2 than regular fresh air [11]. Thus many commercial operations supply pure (100%) CO_2, increasing productivities about ten-fold. However, most *Spirulina* producers in Asia also use bicarbonate and add acetate at the end of the cultivation process. Bubbling air into cultures is another way to deliver CO_2, and is practiced in the production of microalgae in bivalve, shrimp and fish hatcheries, but is expensive for large-scale algae production [12].

3.1.6 Oxygen

Since micro algae produce oxygen during photosynthesis, it is a waste product from the algae and must be evacuated from the growth media efficiently. This process is enhanced by the mixing, used in many open pond designs. In photo bioreactors, oxygen can be produced from photosynthesis at up to 10g O_2/m^3/min [5]. High dissolved oxygen levels inhibit photosynthesis and can lead to algal cell damage.

3.1.7 Water Requirement

The water requirement is important factor for algae production as it plays as a containing media. While evaporative loss of culture volume can seriously affect production due to water availability in arid regions, however, evaporative cooling plays an important and necessary role in maintaining appropriate open pond temperature. Water loss in the range of 0.3– 0.6 cm d-1 [13] is common for open systems and a major problem to algal biomass production [14]. To avoid the water loss, closed systems can be incorporated.

3.2 Biotic Factors

3.2.1 Invasive Species and Predators

Sheehan et al. reports that in open ponds the best algae growth was seen when an invasive species was allowed to take over the pond culture [15]. Research into the issue of invasive species has shown that large clean inoculums are critical to keeping invasive species away [16]. To avoid any problems from such species, algae growth is mostly limited to extremophile algae species.

4. PRODUCTION OF MICRO ALGAE

Though micro algae can grow in their natural habitat as lake, pond, river, coastal streams, and backwater streams however, for production and harvesting biomass, it needs isolated structures. There are some established popular structures used worldwide as i) Open pond (Race way and circular pond), ii) closed systems called as photo bioreactors. Open ponds are open to the environmental conditions and mostly of batch type. Closed systems are of different types like flat panel, tubular, spiral, etc. made up of transparent material, PVC, glass etc. and are semi continuous type.

In India, there are waste water treatment based micro algae production plants, where gasifiers scrubbing water, food industrial waste water, municipal waste water, dairy effluents, etc. are being treated. Some of them are used to harvest biomass and used it for bio fuel production and other are solely for capturing the CO_2, and heavy metals to reduce greenhouse gas production and water pollution respectively. Thus waste water can be used depending on its toxicity.

4.1 Up-Scaling of Micro-Algae Production

Generally selected micro algae strains are stored and maintained in the laboratory and then up scaled step by step. Generally, culture grown in final open pond is avoided to be used as a culture for fresh batch because, micro algae is most likely to be contaminated in the open pond production. The up-scaling of pure strains of selected species of microalgae starts from 10 ml and proceeds through various steps up to the mass production in polyethylene (PE) bags up to 20 l and final growth in tanks (1000 lit and above). Each step involves an increase of the culture volume: when mature, the algal population of a smaller volume is sacrificed to inoculate into larger vessels [16]. Each algal culture sample is monitored every day for cellular growth rates by measuring optical density at 680 nm in a spectrophotometer after calibrating it with a sample of the BG-11 media as blank. The process of up scaling of micro algae production is shown in Fig. 8.2.

Micro Algae Production for Bio Fuel Generation

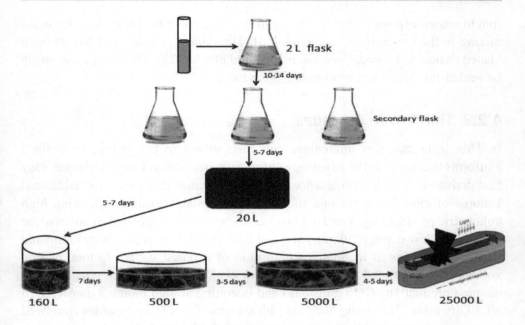

Figure 8.2: Up Scaling of Micro Algae Production

4.2 Microalgae Cultivation

4.2.1 Open Ponds

As an open pond for microalgae growth, it can be characterized into natural sources such as lakes. Shallow big ponds, tanks, circular ponds and Raceway Ponds are the most commonly used systems [17]. The lipid production rate for open ponds is estimated at 39,000 l/ha/year [18].

4.2.2 Raceway Ponds

Raceway ponds are also called intensive ponds or high rate open ponds (HRP). Raceways are the most commonly used artificial systems. These systems are closed loop circulation channels with an oval shape. Inside the channel baffles are installed for guiding the flow of water [5,15]. The depth is usually between 0.2 − 0.3 m [19] so that sun light can pass through the water [15].The ponds are built in compacted earth or in concrete with plastic linings [5, 20]. The algae productivity in raceway ponds can be as much as 10 times higher than in extensive ponds and can reach 14 − 50 g/m^2/d [21,22]. The algae, water and nutrients are kept in circulation and prevented from settling by a continuously working paddle wheel [15]. Aeration of water can be assisted by paddle wheels rotating at around 40 rpm to ensure exposure of media water to the light and atmospheric CO_2 ensuring mixing in the range of 15 – 25 cm/s [15]. The typical paddlewheel has an eight bladed design and a single one can mix a pond of 5 ha [23]. This design can easily be scaled up from one m^3 to many thousands.

4.2.3 Thin Layer Cascades

In Thin layer cascades microalgae culture is grown in the system of inclined platforms that provide the advantages of open systems-direct sun irradiance, easy heat derivation, simple maintenance, and efficient removal of gases with additional features of closed systems-operation at high biomass densities achieving high volumetric productivity. The Thin layer cascades are characterized by microalgae growth at a low depth (<50 mm) and fast flow (0.4-0.5 m/s) of culture compared to raceways ponds. It facilitates a high ratio of exposed surface to total culture volume (>100 l/m) and rapid light and dark cycling of cells which result in high biomass productivity (>30 $g/m^2/day$) and operating at high biomass density, >10 g/L of dry mass [5]. Among important advantages of thin layer cascades compared to raceway ponds are the operation at much higher cell densities, very high daylight productivities.

4.2.4 Circular Ponds

As name suggests, these ponds are circular in shape. Circular ponds are made of a concrete wall and are mixed by a centrally rotating arm. Biomass productivities of 8.5– 21 $g/m^2/d$ were observed [24]. Depth of the pond varies up to 0.4 m.

4.2.5 Closed systems/ Photo bioreactors (PBR)

A. **Flat Plate Photo Bioreactors:** These are basically rectangular in shape (Fig 8.3), the air or CO_2 is provided through bubbling at the bottom, Light source can be provided from outer side or from the sided of the flat plate.

   A transparent material or glass is used for it. The nutrients are provided continuously or as and when needed.

B. **Bubble-Column Photo Bioreactor:** Among the many types of PBRs, the bubble-column type is simple design, easy operating, and energy saving. In this type of photo-bioreactors, aeration is provided at the bottom of the glass or PVC column which helps to provide CO_2 and the circulation of the water. These types of systems are also called vertical tubular PBR [25].

C. Tubular Photo Bioreactor: These are PBRs of cylindrical in shape which can be used in various orientations according to the light source. If sun light is to be used east-waste orientation is usually preferred. Due to tubular shape it has high area to volume ratios, tubular PBRs are known to have high lighting efficiencies [26, 27].

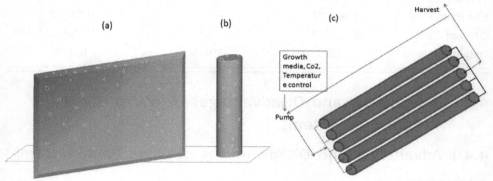

Figure 8.3: (a) Flat Plate Photo Bioreactors, (b) Bubble-Column Photo Bioreactor and (c) Tubular Photo Bioreactor

4.2.6 Hybrid Systems

Hybrid systems are two stage cultivation systems using photo bioreactor and open ponds in combination. The first step is the fast cultivation of biomass in the PBR under controllable conditions. This culture is used as inoculum for large-scale pond operations where the second stage of nutrient stress cultivation is carried out [15].

4.3 Comparison of Cultivation Systems

The cultivation systems show variations in biomass concentration and production depending on different factors. The biomass productivity can be distinguished in area wise productivity [$g/m^2/d$] and volumetric productivity [$g/l/d$]. The typical cell concentrations and biomass productivities of the individual cultivation systems are listed in Table 8.2.

Table 8.2: Cell Concentrations and Biomass Productivities of the Individual Cultivation Systems

Cultivation System	Cell Concentration g/L	Area Wise Biomass productivity (g/m²/day)	Volumetric Productivity (g/l/day)
Raceway pond	0.14-1	2-60	0.05-0.32
Circular pond	-	8.5-21	-

[Table Contd.

Contd. Table]

Cultivation System	Cell Concentration g/L	Area Wise Biomass productivity (g/m²/day)	Volumetric Productivity (g/l/day)
Cascade system	Up to 10	31	-
Tubular PBR	1.5-3	20-114	0.4-1.9
Flat Plate PBR	Upto 80	10-45	0.27-3.8
Column PBR	1.4 – 1.7	93	0.06-0.42
Hybrid Systems	0.3	10-15	0.076

4.4 Advantages and Disadvantages of Micro Algae Production Systems

4.4.1 Advantages and Disadvantages of Open Pond System

Advantages:

i) Open systems are easy to construct and install.

ii) Low operational cost and low maintenance.

iii) Robust design

Disadvantages:

i) More land area and water required as compared to photo bio reactors

ii) Lower algal biomass productivity

iii) In Open pond structures, micro algae are more vulnerable for contamination and environmental conditions than closed one.

iv) In open air pond systems, the culture environment is difficult to control [5].

v) The cultivation is limited by growth parameters including light intensity, pH, dissolved oxygen and temperature [28].

4.4.2 Advantages and Disadvantages of Closed system

Advantages:

In closed structure, biomass can be produced more efficiently than open pond by using nutrient, light intensity, and temperature control.

i) Higher biomass productivity

ii) Due to closed environment potential contamination can be prevented.

iii) Land and water requirement can be reduced more than 10 folds by vertical installation and preventing water evaporation respectively.

iv) Closed systems are more suitable for high value products as it produces quality biomass.
v) Light weight and Portable system

Disadvantages:

i) Installation and operating cost is higher due to equipment used for circulation, aeration, cooling, light intensity, pH control than open ponds
ii) Need higher maintenance than open pond.

5. CONSTRUCTION OF OPEN POND SYSTEMS FOR MICRO ALGAE PRODUCTION

While developing the open pond systems for micro algae production, factors stated below should be considered.

5.1 Operating Factors to be Consider While Development of Open Ponds

5.1.1 Pond Liners

Liners enable a raceway pond to be built in otherwise unsuitable terrain. Liners can be made from different materials such as clay, concrete, asphalt, fiber glass, and HDPE. Clay ponds cannot be cleaned like those utilizing a HDPE liner and can lead to an increase in potential contamination [29]. Despite being costly, liners may be necessary to help mitigate these problems.

5.1.2 Mixing

Gas mixing can be treating as water dynamic for growth of microalgae. Culture mixing is important to allow the algal cells to move in and out of the light zone for photosynthesis. Large 4-ha ponds can be mixed by paddle wheels at speeds of roughly 20–30 cm/sec.

5.1.3 Depth

The depth of a raceway pond is about 0.3 m, which is made of a closed loop recirculation channel. Raceway pond depth is usually 12 to 21 cm [30, 31]. An increase in mixing can lead to a deeper pond being more efficient for algae growth.

5.1.4 Cell Concentration

The cell concentration of the algal culture governs how far the light can penetrate into the media. A high algal cell density also leads to self-shading.

5.2 Construction Design of Raceway and Circular Types Open Ponds

While designing rectangular and circular type open raceway ponds, depending on expected amount of algal biomass, the production of biomass of selected microalgae strain should be known, height of the pond should be kept with extra margin of 20-25% above the 0.2 to 0.3 m, the geared motor should be selected to reduce the rpm to 40-50 rpm or the water flow at the velocity of 15–25 cm/s. For example, for the Capacity of 8 m^3 of the rectangular raceway and circular pond can be constructed as shown in detail drawing of the race and circular pond are given in Fig. 8.4-8.6.

There is provision to fix motor operated paddle wheel in the raceway and circular pond for circulation. The construction includes flooring work, concrete work for baffle, brick work for walls and metal work for paddles and motor support.

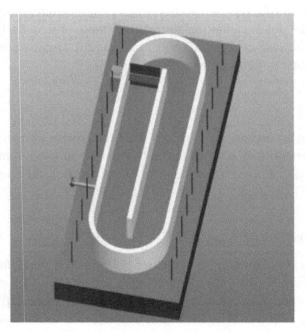

Figure 8.4: CAD Model of the Rectangular Raceway and Circular Pond for Algal Cultivation

Micro Algae Production for Bio Fuel Generation 165

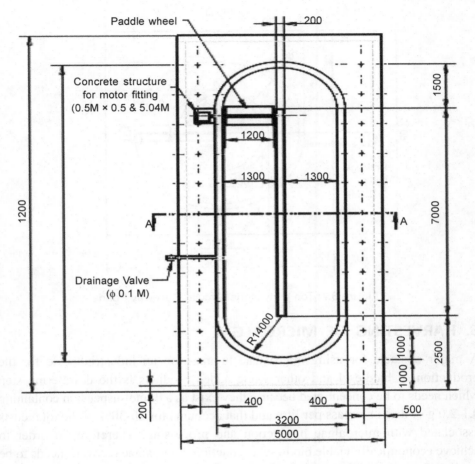

Figure 8.5: Top View Drawing of the Rectangular Raceway Pond

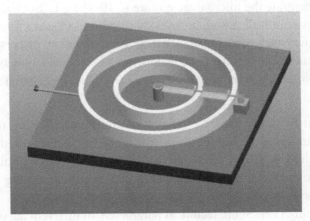

Figure 8.6: The CAD Model of the Circular Pond for Micro Algal Cultivation

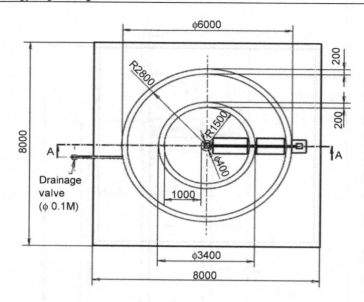

Figure 8.7: Top View of the Circular Raceway Pond

6. HARVESTING OF MICROALGAE

A major obstacle for using microalgae biomass on an industrial-scale for the production of biodiesel and other value added products is the dewatering step which needs to be concentrated because they exist as a dilute suspension containing 0.1-2.0 g of dried biomass per litre and that accounts for 20-30% of the total costs associated with microalgae production and processing. Therefore, in order to achieve economically viable biodiesel production, microalgae recovery needs to be made less costly. Different methods for solid-liquid separation can be employed to dewater/concentrate the microalgae culture to 10-450 g/L. Such methods include sedimentation, vacuum filtration, cross flow filtration, pressure filtration, decanter centrifugation, disc stack centrifugation, dissolved air flotation, dispersed air flotation, micro bubble generation, organic flocculation, inorganic flocculation, bio-flocculation, auto-flocculation and electrolytic coagulation, electrolytic flocculation and electrolytic flotation.

The current methods used for harvesting and concentrating microalgae are characterized in three techniques viz. physical harvesting method, chemical harvesting method and biological harvesting method. Care should be taken to reduce the contamination of chemicals and other materials during harvesting for quality biomass. Approximately 20-30% of the production cost is incurred in the biomass harvesting [11]. Some technical parameters of different harvesting systems are shown in Table 8.3.

Table 8.3: Harvesting Methods of Micro Algae [20, 32]

Process	Main mechanism	Utilities	Dependence on algal strains	Solid conc.	Energy Input
Flocculation: Inorganic Lime	Floc enmeshment	Lime, Mixing	Minor	8-10%	High
Flocculation: Alum	Floc enmeshment and Destabilization	Alum, Mixing	Minor	8-10%	High
Flocculation: Polyelectrolytes	Floc enmeshment and destabilization and bridging	PE, Mixing	Minor	8-10%	Medium
Bio flocculation	Spontaneous flocculation	Pumping, Clarifier	High	1-3%	Low
Centrifugation	Accelerated discrete settling	Power, Equipment	Minor	>10%	High
Cross Flow Filtration	Membrane selfcleaning	Power, Equipment	Minor	2-6%	High
Micro straining	Fabric Straining	Power, Equipment	High	2-4%	Medium

7. DRYING OF HARVESTED ALGAL BIOMASS

Harvested algal biomass still contains 80-90% amount of water [20]. Common methods to remove moisture include sun drying, thermal drying, spray drying, and freeze drying. Spray drying can cause damage to pigments in the algae [33]. After drying micro algal biomass in the powdery form is ready for further conversion into fuels and value added products. Figure 8.8 shows the harvested paste of micro algal biomass and after-drying powder of algal biomass.

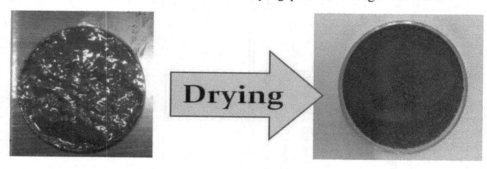

Figure 8.8: Harvested Paste of Micro Algal Biomass and After-Drying Powder of Algal Biomass

8. BIOFUELS DERIVED FROM MICROALGAE

Algal biomass is composed of carbohydrates, lipids, protein (6–52%) [34] and minerals etc. These components can be converted in to bio fuels and value added products [35] by further processing. The bio fuels include biodiesel by trans-esterification process [36], alcohols through fermentation [37], hydrogen and bio-syngas through gasification of algal biomass [38,39,40.] as shown in Fig. 8.9.

9. STATUS, CHALLENGES AND THE WAY FORWARD

The main constrain in bio fuel production from micro algae is the production cost which involve cost of growth media, pre-treatment of substrate used as growth media and its harvesting. There is need of exploring the oil rich micro algae strains dominant in different areas of Indian tropical and sub-tropical conditions. New ways of harvesting by hybrid cultivation of micro algae-bacteria can be further studied. The process of developing low cost media and waste water treatment are needed to be focus along with an enhancement of lipid content in micro algae and media recycling. Different treated crop residue based substrates are adequate in nutrient and produce more biomass than standard growth media.

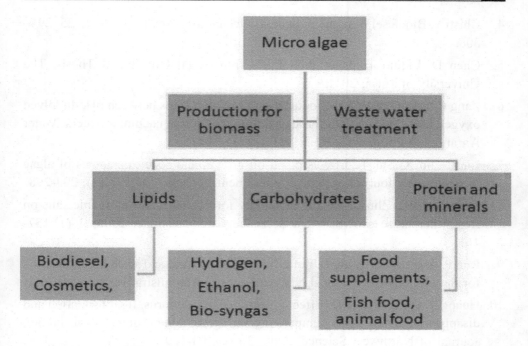

Figure 8.9: Bio Fuels and Value Added Products Derived from Microalgae

This way, carbon capture will be doubled as it utilizes crop residue which is being burn. Many leading research institutes worldwide have understood the importance of micro algae and are working on different aspects of micro algae based bio fuels in the view to make it more viable for bio fuel generation. Considering huge advantages of micro algae biomass as a bio fuel source and its use for other high value products, further research to explore cost reduction possibilities may direct the micro algae as one of the leading arena of renewable energy for India.

REFERENCES

1. Lavens P, Sorgeloos P. Manual on the production and use of live food for aquaculture. 1st ed. Rome: FAO; 1996.
2. Borowitzka AM. Algal biotechnology products and processes-matching science and economics. Journal of Applied Phycology. 1992; 4: 267-279.
3. Munro HGM, Blunt WJ, Dumdei JE, Hickford JHS, Lill ER, Li S, Battershill NC, Duckworth RA. The discovery and development of marine compounds with pharmaceutical potential. Journal of Biotechnology. 1999; 70: 15-25.

4. Chisti Y. Biodiesel from microalgae. Biotechnology Advances. 2007; 25: 294–306.

5. Chen D. Liquid Transportation Fuels from Algal Oils. Ph. *D.* Thesis, The University of Tulsa; 2011.

6. Zang C, Huang S, Wu M. Comparison of relationships between pH, dissolved oxygen and chlorophyll a for aquaculture and non aquaculture waters. Water Air and Soil Pollution. 2011; 219(1-4) :157-174ÿ

7. Liu C, Jin X, Sun L. Effects of ph oil growth and species changes of algae in freshwater. Journal of Agro-Environment Science. 2005; 24(2): 294-298.

8. Liu C, Sun H, Zhu L. Effects of salinity formed with two inorganic salts on freshwater algae growth. Acta Scientiae Circumstantiae. 2006; 26(1):157-161.

9. Ron P. Assessment of algae biofuels resource semand and scale up implication for the US. NNSA (National Nuclear Security Administrative) Report; 2011.

10. Huang Y, Chen M, Liu D. Effect of nitrogen, phosphorus, light formation and disappearance and water temperature on the of blue - green algae bloom. Journal of Northwest Science. 2008; 36 (9): 93-100.

11. Virginia UO. Algae: biofuel of the future? Science Daily. http://www. Sciencedaily. com /releases/2008/08/080818184434.html;2008 [Accessed 02 August 2011].

12. Benemann JR. Microalgae aquaculture feeds. J. Appl. Phycol. 1992; 4: 233–245.

13. Davis R, Fishman D, Frank ED, Wigmosta MS. Renewable diesel from algal lipids: an integrated baseline for cost, emissions, and resource potential from a harmonized model. Argonne National Laboratory, Technical Report: ANL/ESD/12-4, 2012. (www.nrel.gov/docs/fy12osti/55431.pdf).

14. National Research Council. Sustainable Development of Algal Biofuels, http://www.nap. edu/catalog.php?record_id=13437; 2012 [Accessed 09 April 2020].

15. Sheehan J, Dunahay T, Benemann J, Roessler P. A look back at the US Department of energy's aquatic species program-biodiesel from algae. National Renewable Energy Laboratory, Golden, CO; 1998.

16. Ma F, Hanna AM. Biodiesel production: a review. Bioresource Technology. 1999; 70:1-7.

17. Demirbas A. Use of algae as biofuel sources. Energy Conversion and Management. 2010; 51: 2738–2749.

18. Zemke P, Wood B, Dye D. Techno economic analysis of algal photo bioreactors for oil production. NREL-AFOSR joint workshop on algal oil for jet fuel production; 2008.
19. Borowitzka, Michael A. Commercial production of microalgae: ponds, tanks, tubes and fermenters. Journal of Biotechnology 1999.70: 313-321.
20. Brennan L, Owende P. Biofuels from microalgae—A review of technologies for production, processing, and extractions of biofuels and co-products. Renewable and Sustainable Energy Reviews. 2010; 14: 557-577.
21. Alabi AO, Tampier M, Bibeau E. Microalgae technologies & processes for biofuels/bioenergy production in British Columbia. The British Columbia Innovation council Report; 2009.
22. Darzins AL, Pienkos P, Edye L. Current status and potential for algal biofuels production. IEA Bioenergy Task.2010; 39:14-58.
23. Benemann JR, Oswald WJ. Systems and economic analysis of microalgae ponds for conversion of CO_2 to biomass. Final report. US; 1996. DOI: 10.2172/49 33 89.
24. Hosub K, Soonjin H, Haeki S, Jaeki S, Kwangguk A, Chun G. Effects of limiting nutrients and N: P ratios on the phytoplankton growth in a shallow hypertrophic reservoir. hydrobiologia. 2007; 589(1): 255-267.
25. Mubarak M, Shaija A, Prashanth P. Bubble column photobioreactor for *Chlorella pyrenoidosa* cultivation and validating gas hold up and volumetric mass transfer coefficient. Energy Sources Part A: Recovery, Utilization, and Environmental Effects. 2019. DOI: 10.1080/15567036.2019.1680769.
26. Masojídek J, Torzillo G. Mass cultivation of freshwater microalgae. Encyclopedia of Ecology. 2008; 8: 2226-2235. https://doi.org/10.1016/B978-0-12-409548-9.09373-8
27. Tamiya H. Mass culture of algae. Annu. Rev. Plant Physiol. 1957; 8: 309–334.
28. Harun R, Singh M, Forde GM, Danquah MK. Bioprocess engineering of microalgae to produce a variety of consumer products. Renewable and Sustainable Energy Reviews.2010; 14:1037-1047.
29. Lundquist TJ, Woertz IC, Quinn NWT, Benemann JR. A realistic technology and engineering assessment of algae biofuel production. Energy Biosciences Institute, University of California, Berkeley, California; 2010.
30. Vonshak A. Outdoor mass production of spirulina: the basic concept. In: Vonshak A, editors. Spirulina platensis (Arthrospira): Physiology, cell-biology, and biotechnology, London: CRC Press; 1997, p. 79-99.

31. Su Z, Kang R, Shi S, Cong W, Cai Z. An economical device for carbon supplement in large-scale micro-algae production. Bioprocess and Biosystems Engineering. 2008; 31:641-645.

32. Greenwell HC, Laurens LML, Shields RJ, Lovitt RW, Flynn KJ. Placing microalgae on the biofuels priority list: A review of the technological challenges. Journal of the Royal Society Interface. 2010; 7: 703-726.

33. Chen CL, Chang JS, Lee DJ. Dewatering and drying methods for microalgae. Drying Technology. 2015; 33(4): 443-454. DOI: 10.1080/07373937.2014.997881.

34. Viswanathan B. Microalgal application in cosmetics. Microalgae in Health and Disease Prevention. 2018; 3: 317-323.

35. Couteau CL. Microalgal application in cosmetics. Microalgae in Health and Disease Prevention. 2018; 3: 317-323. https://doi.org/10.1016/B978-0-12-811405-6.00015-3.

36. Vasudevan PT, Briggs M. Biodiesel production-current state of the art and challenges. Journal of Industrial Microbiology & Biotechnology. 2008; 35: 421- 430.

37. Dalatony E, Salama MM, Kurade ES, Hassan MB, Oh SHA, Kim SE, Jeon SBH. Utilization of microalgal biofractions for bioethanol, higher alcohols, and biodiesel production: a review. *Energies*. 2017; 10: 2110.

38. Azadia P, Brownbridgea GPE, Mosbacha S, Inderwildib OR, Krafta M. Production of biorenewable hydrogen and syngas via algae gasification: a sensitivity analysis. Energy Procedia. 2014; 61: 2767-2770.

39. Aziz M, Zaini IN. Hydrogen production from algal pathways. In: Meyers R. (eds) Encyclopedia of Sustainability Science and Technology. Springer, New York, NY. 2018. https://doi.org/10.1007/978-1-4939-2493-6_958-1.

40. Abdol GE, Hikmat H. Gasification of algal biomass (*Cladophora glomerata* L.) with $CO_2/H_2O/O_2$ in a circulating fluidized bed. Environmental Technology. 2019; 40(6): 749-755. DOI: 10.1080/09593330.2017.1406538.

CHAPTER - 9

BIOCHAR PRODUCTION FOR ENVIRONMENTAL APPLICATION

Ashish Pawar* and Narayan L. Panwar

Department of Renewable Energy Engineering
College of Technology and Engineering
Maharana Pratap University of Agriculture and Technology,
Udaipur-313001, Rajasthan, India
*Corresponding Author

In India, huge biomass is available from agro waste, forest waste, aquatic waste, municipal solid waste etc. Crop residues have considerable energy potential, if utilized appropriately. Crop residues can be converted in to biochar through thermo-chemical routes; conversion helps in the managing and handling of biomass. Therefore, proper utilization of lignocellulosic crop residues or biomass to make value added product such as biochar by using thermochemical conversion process and followed by activation for the preparation of activated biochar. The application of this biochar to soil, improves the physiochemical characteristics of soil because biochar is rich in organic carbon content, which makes the soil more fertile and acts as the carbon sequestration agent over the long term. The produced biochar is further activated using physical and chemical activation for the preparation of activated biochar. Application of biochar in soil can reduces the emission of carbon dioxide gases into the atmosphere while simultaneously increasing soil fertility. Activated Biochar is having ability to minimize the environmental problems like accretion of agricultural residues or waste, air pollution as well as water pollution. The different production technologies of biochar and its applications are discussed in present book chapter.

1. INTRODUCTION

In India, there are 686 MT gross residues are available from agriculture crops in which 23.5 MT signifies the surplus residues [1]. Some peasants are burning these agro wastes in open land which creates environment pollution and significantly affect living organisms. So, proper utilization of surplus lignocellulosic crop residues or biomass to make value added product such as biochar by using thermochemical conversion process like pyrolysis or carbonization became the urgent need for the modern society. Biochar (BC) is having ability to minimize the environmental problems like accretion of agricultural residues or waste, air pollution as well as water pollution. Generally, any biomass or crop residue mainly composed of three different constitutes, which are hemicelluloses, cellulose and lignin. Among the three, lignin found effective component in a biomass for the production of biochar which is more suitable for adsorption process [4]. Apart from being, crop residues or biomass obtained from agriculture waste has been become a promising type of precursor due to its low cost and abundantly availability for the preparation of biochar. Any agriculture waste having a high carbon content and low inorganic content may be preferred for the production of biochar. [5]. In addition, instead of using fossilresources such as coal, biochar obtained from lignocellulosic biomass or crop residues helps in reduces the global warming effect. Therefore, circulation of carbon into atmosphere and pollutant removal process takes place simultaneously and it becomes carbon neutral cycle [6]. The biochar produced from agro waste is shown in Fig. 9.1

Figure 9.1: Biochar Produced from Agrowaste [6]

The present chapter focused on brief introduction of definition of biochar, present status in Indian context, various technologies for biochar production, physicochemical properties, environmental applications, and finally various advantages and disadvantages of biochar.

2. BIOCHAR

Biochar is an organic carbonaceous material obtained during the thermochemical conversion of biomass mainly via carbonization process. In case of carbonization process organic biomass or any waste material is heated at temperature range from 400 to 600 °C in absence of air or oxygen gives three primary end products solid (biochar), liquid (Bio-oil) and gaseous fuel [7]. A solid carbonaceous material obtained during the carbonization of biomass simply known as biochar. As biochar is carbon rich, environment friendly, and cost effective, therefore it can be used for various applications for ex. as a soil remediation, energy production, waste water treatment, fuel cells and as an electrode in supercapacitor for energy storage application. The organic part of biochar is mainly composed of carbon, while inorganic portion made up of calcium, magnesium, potassium and other inorganic constituents. The properties and structure of biochar is mainly depending on the type of feedstock, operating condition (Carbonization temperature, heating rate and residence time) etc. [4].

3. STATUS IN INDIAN CONTEXT

As per Rajagopal et al. [8] estimation, the India is having annual potential to produce biochar about 75.2 Mt to 219.33 Mt from fast and slow pyrolysis using surplus agricultural residue. According to author, the slow pyrolysis process being a more efficient process for the production of biochar and produces 78.2 Mt of biochar from only agriculture waste. In India, there are 686 MT gross residues are available from agriculture crops in which 23.5 MT signifies the surplus residues [1]. According to Ministry of New and Renewable Energy, Govt. of India, in north-east region of India is having a potential to produce 37 million tons of agriculture and forest biomass. If only 1 per cent of biomass is converted into value added organic carbon rich biochar, then 74 thousand tons of carbon may be sequestrating annually. In India, total 31 terra joule of energy is produced from 1 per cent conversion of biomass into biochar, 1300 and 900 tons of bio-oil and biogas respectively [9].

4. BIOCHAR PRODUCTION TECHNOLOGIES

Biochar is a carbon rich, fine-grained, and porous organic product which produced when biomass has been subjected to thermo chemical conversion process (Slow pyrolysis) at temperatures range between 350-600°C in an environment with limited or in absence oxygen [10]. There are so many techniques are available for the production of biochar by optimizing the thermo chemical conversion routes, which are illustrated in Fig. 9.2.

The yield, physiochemical composition of biochar is mainly depending on various biochar conversion technologies, its operating condition and available feedstock. Owing to this sense, Xie et al. [11] reported that maximum biochar yield (from 15 to 35 %) was obtained by keeping the biomass at longer residence time (up to 4 h) with a operating temperature about 500°C, and on the other hand maximum bio-oil yield was recorded at very less residence time (up to 2 s).

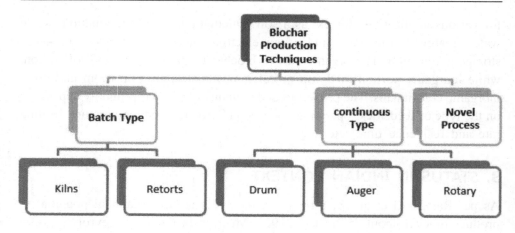

Figure 9.2: Biochar Production Techniques

4.1 Batch Type

In batch type process, kilns it may be earthen and mound, metal, brick, concrete kiln etc. and another one is retorts are used for biochar production by using a different kind of feed stocks such as woody biomass, forest waste and agricultural waste. Anciently, batch type process was used for biochar production in rural areas due to its low operational and constructional cost. Although, the biochar yield in such processes varies from 12.5-30% [12], but it takes more residence time (hours to days) for carbonization process. Pennise at al. [13] produced charcoal by using brick kiln (Fig. 9.3), having a capacity of 20,000 kg of woody biomass and recorded maximum charcoal yield (68.9%), the resulting charcoal having carbon content about 85.7% and its heating value was about 29.20 KJ/kg.

4.2 Continuous Type

In case of continuous mode of pyrolysis, biochar is obtained through drum type pyrolyzer, auger reactor method, and using rotary kilns. Now a day, continuous process for biochar production is usually adapted in both industrial and commercial sector because of its maximum conversion efficiency, energy efficient technique and resulting biochar having a good quality. The main advantage of this process is continuous biochar production take place along with a greater flexibility towards the biomass feedstock. The maximum biochar yield was recorded in the range of 25 to 35% [15]. Drum type pyrolyzer is also effective for the continuous production of biochar, horizontally mounted drum was rotated and external heat is provided about 400-450°C to achieve the carbonization process. The continuous loading of raw material and due to rotary rotation, the movement of biomass takes place

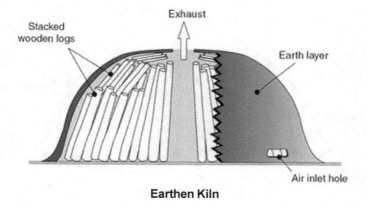

Earthen Kiln

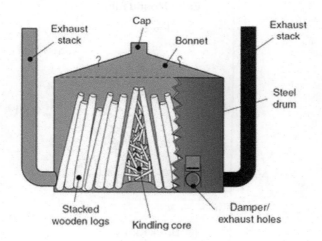

Metal Kiln

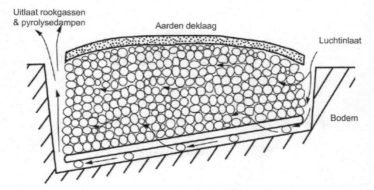

Pit Kiln

Figure 9.3(a): Batch Type Kilns for Biochar Production Used Traditionally [14]

178 Bioenergy Engineering

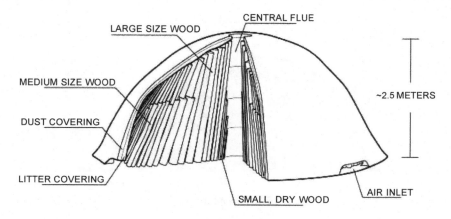

Earth Mound Kiln

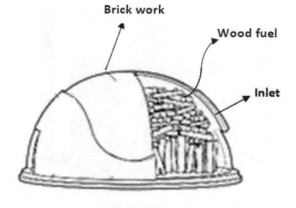

Brick Kiln

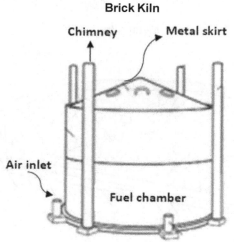

TPI* Transportable Metal Klin

Figure 9.3(b): Batch Type Kilns for Biochar Production Used Traditionally [14]

inside the reactor, which causes uniform distribution of heat through the biomass. The rotary drum type pyrolyzer gives maximum efficiency about 50% and almost 90% heat could be recovered during the carbonization process [16].

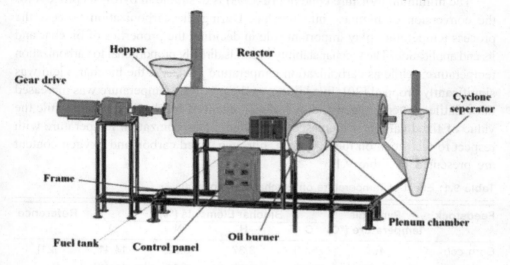

Figure 9.4: Continuous Biochar Production System [16]

4.3 Novel Process

Flash carbonization process is also known as novel process, in which biomass is quickly and very efficiently converted into biochar. The operating pressure play significant role in carbonization process, generally pressure was kept between 1-2 MPa by using compressed air [17]. The maximum biochar yield is obtained from novel process was ranged between 40-50 %, having almost 70 to 80 % fixed carbon content [18]. Evans et al. produced biochar from corncob and nut shell biomass using a novel process of flash carbonization. The selected biomass sample was placed in packed bed at moderate temperature with a pressure of 1-2 MPa. Due to maintaining the high pressure inside the reactor the system increases the biochar yield.

5. FACTORS AFFECTING BIOCHAR PRODUCTION

The performance of biochar production technology is mainly depending on the type of feedstock, particle size, moisture content present in biomass, carbonization temperature, residence time, heating rate, etc. The biomass is mainly composed of cellulose, hemicelluloses and lignin. The different operating conditions are significantly affected on the physicochemical properties and biochar yield. Moisture content present in biomass affects on the reaction chemistry and also the properties as well as yield of end product [19].

The minimum moisture content (10-15%) is desirable in pyrolysis process for the conversion of biomass into biochar. During the carbonization process the process temperature play important role in deciding the properties of biochar and its end application. The biochar stability in soil is directly proportional to carbonization temperature; while as carbonization temperature increased the biochar, yield was significantly dropped [20]. In addition, as the pyrolysis temperature was increased the volatile matter, oxygen and hydrogen content of biochar decreased, while the value of fixed carbon is increased. The effect of carbonization temperature with respect to feedstock on the hydrogen, nitrogen, fixed carbon and oxygen content are presented in Table 9.1.

Table 9.1: Effect of Temperature on Biochar Composition

Feedstock	Pyrolysis temperature (°C)	Biochar Elements (%)				Reference
		C	H	N	O	
Corn cob	400	75.23	3.37	0.82	14.11	[21]
	450	77.84	2.95	0.86	11.45	
	500	80.85	2.5	0.97	8.87	
	550	82.62	2.25	0.84	7.43	
Rapeseed	400	57.95	3.43	5.43	33.16	[22]
	450	59.77	2.36	5.12	32.75	
	500	61.98	1.92	4.12	31.78	
	550	67.29	1.75	4.35	26.21	
Safflower seed	400	68.76	4.07	3.77	23.49	[23]
	450	70.43	3.49	3.69	22.39	
	500	71.37	2.96	3.91	21.76	
	550	72.96	2.67	3.74	20.63	
	600	73.72	2.34	3.84	20.10	
Conocrpus waste	200	64.20	3.96	0.69	26.60	[24]
	400	76.80	2.83	0.87	14.20	
	600	82.90	1.28	0.71	6.60	
	800	85.00	0.62	0.90	4.90	

6. PHYSICOCHEMICAL PROPERTIES OF BIOCHAR

The partial or complete decomposition of biomass at elevated temperature in absence of air resulted in primary solid end product known as biochar or charcoal. The physicochemical properties of biochar significantly vary according to pyrolysis condition and type of feedstock. Generally, slow pyrolysis process widely adopted for getting a higher yield of biochar (35 %) followed by gas (35%) and biooil (30 wt %) at a pyrolysis temperature (300-800°C) for longer duration (maybe hours). Wang et al. (2015) [25], summarized the elemental and physicochemical composition of biochar from pyrolysis of different agrowaste at different heating rate and pyrolyzing temperatures (300-600°C) as, carbon-containing (60-80%), N (0.3-3.10), P (0.03-0.3), K (0.02-1.21), O/C atomic ratio (0.1-0.4), H/C atomic ratio (0.3-0.8), pH (7-10), BET surface area (0.1-16 m^2/g) and pore volume (0.002-0.02 ccg^{-1}) respectively. All the produced biochar was rich in organic carbon, which was due to as pyrolysis temperature increased the organic carbon content in biochar also improved, while oxygen and hydrogen content were significantly reduced. Also decrease in H/C and O/C ratio of all biochar at higher temperature indicated loss of oxygen-containing functional groups and formed graphic as well as aromatic structure. pH of biochar may be influenced by pyrolysis temperature, as pyrolysis temperature increased more alkaline cations (Ca, Mg and k) accumulate on biochar surface which causes increase in pH. The surface area, pore-volume, i.e. the microscopic surface structure of biochar after pyrolysis were showed a good potential ability for adsorption and filtration of organic and inorganic pollutants [26]. The higher heating value of biochar (15-30 MJ/kg), varies according to pyrolysis condition and feedstock [27]. The maximum heating values of biochar become an attractive alternative to coal in fuel application.

Recently, Siddiqui et al (2019) [28] characterized biochar produced from slow pyrolysis of waste pomegranate peel as biomass at a temperature between 300-600 °C, by keeping reaction time from 20-60 min and recorded maximum yield of biochar 54.9 % at 300 °C within a 20 min of residence time. The produced biochar showed higher heating value (14.61 MJ/kg), carbon content (44.5%), oxygen (37.8%), and hydrogen (5.28%) respectively. In addition, SEM analysis of biochar clearly indicated that an increase in pore size and surface area of biochar could be used for environmental application, agriculture application and also for wastewater treatment. The physicochemical properties of biochar at different operating conditions are presented in Table 9.2.

7. APPLICATIONS OF BIOCHAR

Biochar having various applications in rural areas, commercial sector, and industrial sector such as it acts as soil amendment, for waste water treatment purpose, energy production, in cosmetics, and as a substitute for paints and coloring application as shown in Fig. 9.5.

Table 9.2: Physiochemical Properties of Biochar at Different Operating Conditions.

S. No	Feedstock	Pyrolysis conditions		Biochar yield %	Proximate analysis (%)				Ultimate analysis (%)				HHV (MJ/k)	BET surface area (m²/g)	Pore volume (m³/g)	Porosity (%)	O/C ratio	H/C ratio	pH	References
		Temp. (°C)	Heating rate/ residence time		V.M.	F.C.	Ash	M.C.	C	H	N	O								
1	Hinoki cypress	350	10-15 °C/min; 1 h	32.7	41.17	58.52	0.32	-	75.74	5.29	0.22	18.75	29.47	2.40	-	-	-	-	7.95	[29]
2	Avicennia marina (uncontaminated)	300	10 °C/min	60.2	52.70	36.09	7.75	3.46	55.07	5.69	0.67	38.35	27.20	-	-	-	-	-	-	[30]
3	Medium density fibreboard (lignocellulosic biomass)	450	15 °C/min; 30 min	27.6	4.1	88.6	5.0	1.5	80.34	2.91	4.54	7.20	-	-	-	-	0.07	0.43	-	[31]
4	Biomass pellets	400	10-15 °C/min	35-40	27	60	8	5	61.55	1.87	1.22	-	27	300	0.15	0.70	-	-	-	[32]
5	Peanut stalk	600	10 °C/min; 2 h	-	-	-	19.3	-	53.2	3.67	1.87	23.4	-	279	-	-	-	-	6.84	[33]

Biochar Production for Environmental Application

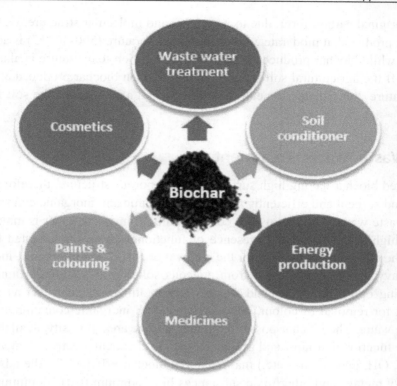

Figure 9.5: Various Applications of Biochar

7.1 Soil Conditioner

Biochar is carbon rich organic material which acts as a soil amendment. Biochar produced from pyrolysis of biomass shows significant properties in soil such as in case of acidic soil; it helps for enhancing the pH, water holding capacity, and cation exchange capacity of soil [34]. The application of biochar in specific proportion can increase the organic matter content in soil and the soil fertility. The proper mixing of biochar in soil can significantly improves the soil structure, porosity, texture, density and particle size distribution etc. Due to proper dose of biochar in soil, the positive changes occurred in case of plant growth, crop yield, reduced soil acidity, improved water quality, increased water holding capacity, reduced leaching of nutrients, and which results in minimum requirement of irrigation. Biochar play significant role in carbon sequestration in which biochar have an ability to capture the carbon and stored in soil to prevent its emission into atmosphere. The biochar stay in soil for longer duration, which causes a resistance to biological and chemical degradation, which results in increases in carbon stock in soil. Biochar is chemically and biologically strong and stable material as compared to the original carbon form, due to its origins and molecular structure. Generally, biochar produced at moderate carbonization temperature (300-400°C) is acidic in nature, while biochar produced at higher carbonization temperature is alkaline in nature. If the agricultural soil is acidic in nature, then biochar produced at higher temperature 700°C helps to neutralize the soil and also enhance the soil fertility [35].

7.2 Waste Water Treatment

Activated biochar having high surface area and porous structure, therefore it acts as a good sorbent and efficiently removes the organic and inorganic contaminates from waste water. The removal of contaminants from waste water is mainly due to the high surface area and presence of functional group on activated biochar [36]. The utilization of biochar for the removal or adsorption of organic, inorganic and heavy metal contaminants from aqueous solution is become an innovative promising option for water and waste water treatment. Biochar act as a super sorbent for removal of colour, odour, organic, and inorganic contaminants from potable water. The biochar possesses higher surface area, porosity, availability of surface functional groups, and presence of some chemically active elements (e.g. COOH, OH, and ketones etc.) that make the biochar efficient for the adsorption of heavy metals, and other toxic substances like cadmium (Cd), aluminium (Al), manganese (Mn), lead (Pb), copper (Cu) from both contaminated water and soil. The biochar possesses negative charge on their surface due to the availability of oxygenated functional groups. Consequently, biochar is used as a cost-effective adsorbent for the removal of organic, inorganic contaminants, and heavy metals from waste water [37]. Therefore, appropriate use of biochar or biochar based nano composites in aqueous media creates a good platform for the removal of contaminants.

7.3 Energy Production

Presently, the energy consumption rate is increasing due to rapid growth in population. Therefore, overdependence on fossil fuel is rising beyond the limit which causes harmful effect on environment, there is need to store clean and green energy by using an alternative material from waste for achieving the social and economic development. Biochar produced from waste have several applications in energy and commercial sectors [38]. Biochar having high calorific value due to maximum carbon content, so biochar can be utilized for making pellets, further it can be used as combustion fuel or a substitute for fossil fuel. According to various literatures, biochar is act as an electrode can also use in supercapacitors, mainly used for energy storage for hybrid vehicles [39,40]. There are various energy storage devices like fuel cells, batteries (lithium ion) and supercapacitors etc. Among them supercapacitor device is widely used energy storage device in electric vehicles and wireless terminals due its outstanding electrical performance. Therefore, there is need to carry out more research for the development of reliable and promising electrode material to improve its electrochemical performance as well as charging and discharging performance. Super capacitors are normally used nonrenewable material like graphite and other inorganic metal having a higher cost. In contrast, biochar produced from biomass is renewable, sustainable, cost effective, and eco-friendly material which shows inherent properties in terms of its mechanical strength and can be used along with another material. Therefore, biochar-based material is very useful to develop a super capacitor which shows the high-power density and provide inherent batteries with a very high electrochemical capacity [41].

7.4 Cosmetics

Activated biochar can also be used as an additive in skin creams, soaps, and as therapeutic bath additives.

7.5 Paints and Colouring

Sometimes, biochar also used as an additive in industrial paints and colouring.

8. ADVANTAGES AND DISADVANTAGES OF BIOCHAR

The lists of potential benefits or advantages of biochar, on the other hand there are few drawbacks have been found by various researchers [17, 42] are shown in Fig. 9.6.

8.1 Advantages of Biochar

- Biochar is easy to recycle, act as a co-catalyst and low-cost material for syngas cleaning and biodiesel production.
- Biochar as a sustainable resource for application in soil which reduces the greenhouse gas emission, nutrient losses and demand of fertilizers.
- Mostly use in fuel cell for energy storage.

- Due to oxygenated functional groups biochar act as a good sorbent for removal of contaminants.
- Due to higher recyclability biochar can be used in hydrogen storage application.
- Biochar material is also used as a precursor material for preparation of activated biochar.

8.2 Disadvantages of Biochar

- Biochar as a catalyst possess lower efficiency as well as abrasive resistance than commercial catalyst.
- Sometime due to lack of physicochemical properties biochar adsorb only possible heavy metals.
- Due to high ash content in biochar possess low voltage and power output for application in fuel cell.
- It requires surface treatment for further energy storage application.
- Required much more cost and patience for the preparation of activated biochar.

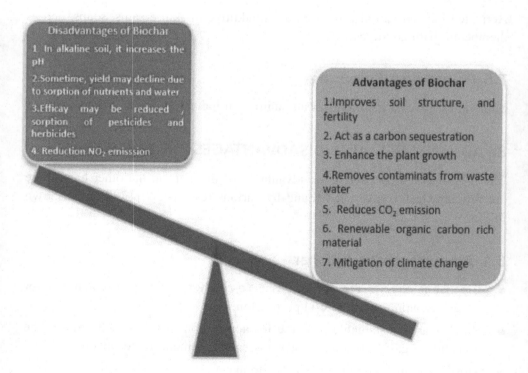

Figure 9.6: Advantages and Disadvantages of Biochar

9. ECONOMICS OF BIOCHAR

Generally, biochar is produced from biomass through slow pyrolysis of fast pyrolysis. Therefore, the production cost of biochar is mainly comprised of handling of raw material, pre-treatment, pyrolysis, electricity requirement, heat energy needed for carbonization etc. [42] slow and fast pyrolysis are mainly effect economics of the production of biochar. The pyrolysis process becomes more profitable when we can reduce the cost of feedstock. When the price of biochar exceeds Rs. 18, 432 t^{-1}, then slow pyrolysis process becomes more profitable. In case of demand of electricity, fast pyrolysis becomes more profitable, when electricity prices exceed above Rs. 8616 MWh^{-1}. On the other hand, slow pyrolysis required more prices above Rs. 22, 778 MWh^{-1}. According to Galinato SP et al. [43] farmer will get more profit after application of biochar to soil, only when the market value of biochar is low (i.e. when biochar value lowers than Rs. 900 and Rs. 7492/MT, when the price of carbon offset is Rs. 75 and 2322/MT CO_2, respectively). In India, the market value of biochar in the range of Rs.150-200/kg and the production cost required to produce a kg of biochar is in the range of Rs.40-50.

10. CONCLUSIONS

In countryside area, crop residues are mainly used as an animal feed and remaining residue i.e. handling of surplus residue is became one of the important threats in front of some peasants. The burning of surplus crop residues in open field creates environmental pollution. Therefore, conversion of waste into value added product i.e. biochar via thermo chemical conversion process may circumvents the problems related to environment pollution, creates employment for people living in rural area, and future economic opportunities. The biochar produced at high carbonization temperature (400-600°C) shows high fixed carbon content, higher porosity, larger surface area, good physicochemical properties, availability of functional groups and nutrients. Therefore, it has extensive applications in the environmental section like as a soil conditioner, waste water treatment, energy generation, fuel cells, cosmetics etc.

REFERENCES

1. Hiloidhari M, Das D, Baruah DC. Bioenergy potential from crop residue biomass in India. Renewable and Sustainable Energy Reviews. 2014; 32: 504-12.
2. Hiremath MN, Shivayogimath CB, Shivalingappa SN. Preparation and characterization of granular activated carbon from corn cob by KOH

activation. International Journal of Research in Chemistry and Environment. 2012; 2(3): 84-87.

3. Zanzi R, Bai X, Capdevila P, Bjornbom E. Pyrolysis of Biomass in Presence of Steam for Preparation of Actovated Carbon. Liquid and Gaseous, 2001.

4. Carrott PJM, Carrott MR. Lignin–from natural adsorbent to activated carbon: a review. Bioresource Technology. 2007; 98(12): 2301-2312.

5. Tsai WT, Chang CY, Lee SL. Preparation and characterization of activated carbons from corn cob. Carbon. 1997; 35:1198–200.

6. Nor NM, Lau LC, Lee KT, Mohamed AR. Synthesis of activated carbon from lignocellulosic biomass and its applications in air pollution control—a review. Journal of Environmental Chemical Engineering. 2013; 1(4): 658-666.

7. Herath HMSK, Camps-Arbestain M, Hedley M, Van Hale R, Kaal J. Fate of biochar in chemically-and physically-defined soil organic carbon pools. Organic geochemistry. 2014; 73: 35-46.

8. Rajagopal V. Prospects of biochar in climate change mitigation in indian agriculture - an analysis. International Journal of Agriculture Sciences, ISSN: 0975-3710. 2018.

9. Srinivasarao C, Gopinath KA, Venkatesh G, Dubey AK, Wakudkar H, Purakayastha TJ., Mandal S. Use of biochar for soil health management and greenhouse gas mitigation in India: Potential and constraints. Central Research Institute for Dryland Agriculture. Hyderabad, Andhra Pradesh 2013.

10. Amonette JE, Joseph S. Characteristics of biochar: micro chemical properties, biochar for environmental management: Science and Technology. 2009.

11. Xie T, Reddy KR, Wang C, Yargicoglu E, Spokas K. Characteristics and applications of biochar for environmental remediation: a review. Crit Rev Environ Sci Technol. 2015; 45(9): 939–969

12. Kammen DM, Lew DJ. Review of technologies for the production and use of charcoal. Renewable and appropriate energy laboratory report. http://rael.berkeley.edu /old_drupal/sites/default/fles/old-site-fles/2005/Kammen-Lew-charcoal2005.pdf 2005.

13. Pennise DM, Smith KR, Kithinji JP, Rezende ME, Raad TJ, Zhang J, Fan C. Emissions of greenhouse gases and other airborne pollutants from charcoal making in Kenya and Brazil. J Geophys Res Atmos. 2001; 106: 24143–24155

14. Shamim MIA, Uddin N, Hossain SAAM, Ruhul A, Hiemstra T. Production of biochar for soil application: A comparative study of three kiln models. Pedosphere. 2015; 25: 696-702.

15. Duku MH, Gu S, Ben HE. Biochar production potential in Ghana: a review. Renew Sustain Energy Rev. 2011; 15: 3539–3551

16. Pawar A, and Panwar NL. Experimental investigation on biochar from groundnut shell in a continuous production system. Biomass Conversion and Biorefinery. 2020; 1-11.

17. Wade SR, Nunoura T, Antal MJ. Studies of the fash carbonization process. 2. Violent ignition behavior of pressurized packed beds of biomass: a factorial study. Ind Eng Chem Res. 2006; 45: 3512–3519.

18. Evans RJ. The relation of pyrolysis processes to charcoal chemical and physical properties. National Renewable Energy Laboratory, Diakses Pada. 2008; 18.

19. Mohan D, Pittman CU, Steele PH. Pyrolysis of wood/biomass for Bio-oil: a critical review. Energy and Fuels. 2006; 20: 848-889.

20. Brown R. Biochar production technology. Chapter 8 in J. Lehmann and S. Joseph (eds) Biochar for Environmental Management. Earthscan, London. 2009.

21. Manya JJ, Roca FX, Perales JF. TGA study examining the effect of pressure and peak temperature on biochar yield during pyrolysis of two-phase olive mill waste. Journal of Analytical and Applied Pyrolysis. 2013; 103: 86-95.

22. Manya JJ, Ortigosa MA, Laguarta S, Manso JA. Experimental study on the effect of pyrolysis pressure, peak temperature and particle size on the potential stability of vine shoots-derived biochar. Fuel. 2014; 133: 163-172.

23. Al-Wabel MI, Al-Omran A, El-Naggar AH, Nadeem M, Usman A. Pyrolysis temperature induced changes in characteristics and chemical composition of biochar produced from conocarpus wastes. Bioresour Technol. 2013, 131: 374–379

24. Wang K, Peng N, Lu G, Dang Z. Effects of pyrolysis temperature and holding time on physicochemical properties of swine-manure-derived biochar. Waste and Biomass Valorization. 2018; 1-12. doi.org/10.1007/s12649-018-0435-2

25. Wang S, Gao B, Zimmerman AR, Li Y, Ma L, Harris WG, Migliaccio KW. Physicochemical and sorptive properties of biochars derived from woody and herbaceous biomass. Chemosphere. 2015, 134: 257-262.

26. Ahmad M, Rajapaksha AU, Lim JE, Zhang M, Bolan N, Mohan D. Biochar as a sorbent for contaminant management in soil and water: a review. Chemosphere. 2014; 99: 19–33

27. Tag AT, Duman G, Ucar S, Yanik J. Effects of feedstock type and pyrolysis temperature on potential applications of biochar. Journal of Analytical and Applied Pyrolysis. 2016; 120: 200-206.
28. Siddiqui MTH, Nizamuddin S, Mubarak NM, Shirin K, Aijaz M, Hussain M, Baloch HA. Characterization and process optimization of biochar produced using novel biomass, waste pomegranate peel: a response surface methodology approach. Waste and Biomass Valorization. 2019; 10(3): 521-532.
29. Yu S, Park J, Kim M, Ryu C, Park J. Characterization of biochar and byproducts from slow pyrolysis of hinoki cypress. Bioresource Technology Reports.
30. He J, Strezov V, Kumar R, Weldekidan H, Jahan S, Dastjerdi BH, Kan T. Pyrolysis of heavy metal contaminated Avicennia marina biomass from phytoremediation: characterisation of biomass and pyrolysis products. Journal of Cleaner Production, 2019; 234: 1235-1245.
31. Haeldermans T, Claesen J, Maggen J, Carleer R, Yperman J, Adriaensens P, Schreurs S. Microwave assisted and conventional pyrolysis of MDF–characterization of the produced biochars. Journal of Analytical and Applied Pyrolysis. 2019; 138: 218-230.
32. Husain Z, Ansari K-B, Chatake VS, Urunkar Y, Pandit AB, Joshi JB. Valorisation of biomass pellets to renewable fuel and chemicals using pyrolysis: characterisation of pyrolysis products and its application. Indian Chemical Engineer. 2019; 1-14.
33. Tan W, Wang L, Yu H, Zhang H, Zhang X, Jia Y, Xi B. Accelerated microbial reduction of azo dye by using biochar from iron-rich-biomass pyrolysis. Materials. 2019; 12(7): 1079.
34. Tan XF, Liu SB, Liu YG, Gu YL, Zeng GM, Hu XJ, Jiang LH. Biochar as potential sustainable precursors for activated carbon production: multiple applications in environmental protection and energy storage. Bioresource Technology. 2017; 227, 359-372.
35. Rodriguez-Reinoso F, Schuth F, Sing KSW, Weitkamp J. Handbook of Porous Solids. Wiley-VCH Verlag GmbH, Weinheim. 2002; 3 (1).
36. Rodriguez-Reinoso F, Molina-Sabio M. Advanced Colloid Interface Science. 1998; 76–77, 271–294.
37. Chen Y, Zhu Y, Wang Z, Li Y, Wang L, Ding L, Guo Y. Application studies of activated carbon derived from rice husks produced by chemical-thermal process—a review. Advances in Colloid and Interface Science. 2011; 163(1): 39-52.

38. Lozano-Castello D, Lillo-Rodenas MA, Cazorla-Amoros D, Linares-Solano A. Carbon. 2001; 39: 741–749.

39. Smebye A, Alling V, Vogt RD, Gadmar TC, Mulder J, Cornelissen G, Hale SE. Biochar amendment to soil changes dissolved organic matter content and composition. Chemosphere. 2016; 142: 100-105.

40. Rajapaksha AU, Chen SS, Tsang DC, Zhang M, Vithanage M, Mandal S, Ok YS. Engineered/designer biochar for contaminant removal/immobilization from soil and water: potential and implication of biochar modification. Chemosphere. 2016; 148: 276-291.

41. Tan XF, Liu SB, Liu YG, Gu YL, Zeng GM, Hu XJ, Jiang LH. Biochar as potential sustainable precursors for activated carbon production: multiple applications in environmental protection and energy storage. Bioresource Technology. 2017; 227: 359-372.

42. Lehmann J, Joseph S. Biochar for environmental management: science, technology and implementation. Routledge. 2015.

43. Galinato SP, Yoder JK, Granatstein D. The economic value of biochar in crop production and carbon sequestration. Energy Policy. 2011; 39(10): 6344-6350.

CHAPTER - 10

ADVANCEMENT IN IMPROVED BIOMASS COOKSTOVE AND ITS CURRENT STATUS IN INDIA

Himanshu Kumar[1]*, Amit Ranjan Verma[1], Swapna S. Sahoo[1] and Narayan L. Panwar[2]

[1]*Centre for Rural Development and Technology*
Indian Institute of Technology Delhi, HauzKhas, New Delhi-110016, India
[2]*Department of Renewable Energy Engineering*
College of Technology and Engineering
Maharana Pratap University of Agriculture and Technology,
Udaipur-313001, Rajasthan, India
*Corresponding Author

In India, around 60-70% of the total population resides in rural areas. Of the total, about 85% are dependent on locally available biomass to meet their cooking energy and water heating demand. The burning of biomasses in an open fire or a simple cookstove is inefficient as it wasted around 90-95% of its energy content. Meanwhile, household air pollution is another serious problem associated with the above cooking practices and caused the premature death of about 0.57-0.78 million people in India annually. The incomplete combustion of solid biomass in such stoves leads to emit carbon monoxide and particulate matter caused chronic disease among the women and children. Though, India is blessed with a considerable potential of biomasses of around 234 MT as surplus biomass. The efficient utilization of these biomasses could rid of the energy needs of rural households through a clean cooking practice. To curb the household air pollution and to improve thermal efficiency, a variety of technological intervention on design, development and evaluation of improved biomass cookstove has been done so far. However, the higher upfront cost of these cookstoves, lack of knowledge about the benefits of such stove among households, lack in coordination among implementation agencies, and design of cookstove based on specific user choices are the primary cause of low acceptance of these newly evolved technologies. Thus, the present chapter focused on how to enable the adoption of improved cookstove in daily life, the improved cookstove classification, recent advancement in technology, potential of ICS to mitigate GHG emissions, support of policy framework in the country, and barriers in technology dissemination and adoption.

1. INTRODUCTION

The biomass is a renewable matter available locally at free of cost in India. The current potential of around 234 MT as surplus biomass would require efficient technologies to harness its maximum energy [1]. In India, about 60-70% of the overall population lives in rural parts [2] and of the total rural population, about 85% are dependent on locally available biomass to meet their cooking and water heating demand [3].The biomass and coal contribute a share of 75-95% of the primary energy needs in rural households [4]. The use of biomass to get cooking and water heating energy requirements through a traditional/ simple stove (shown in Fig. 10.1) in rural India is an age-old practice [5].

Figure 10.1: A Typical Traditional Cookstove Used in Rural Households [6]

Presently, around 2.8 billion people rely on traditional or simple cookstove globally [7], whereas, in India, around 826 million used simple cookstove to burn solid biomass and coal [8]. It has estimated that around 20% of rural households will depend on biomass even by the end of the year 2047 [9]. The traditional cookstoves have characterized by a lower thermal efficiency (5-10%) andhigher emission of household pollutants due to incomplete combustion of the fuels [5, 10]. The higher household pollutants inside the dwelling caused an annual premature death of around 1.6 to 4.3 million globally and 0.57 to 0.78 million people in India

[7, 11, 12, 13]. To curb the household air pollution and to improve thermal efficiency various improved biomass cookstove has been developed [14]. An improved cookstove (ICS) is a thermo chemical device design with certain scientific principles to improve better biomass consumption through improved combustion, low emission of pollutants, and enhanced thermal efficiency [15]. The ICS can save a significant amount of carbon dioxide emission concerning traditional/ simple cookstove and even better in emission than LPG stove [10]. Whereas, the surplus biomasses in the country could be used in such cookstoves to save the time of cooking, fuel cost, reduce the drudgery of women in fuel wood collection etc. However, the acceptances of improved cookstoves are less than one per cent, and only 14% of Indian households are cognizant of ICS [9]. Thus, the current chapter focused on describing the ICS, its recent advancement, potential to mitigate GHG emissions, the policy framework to disseminate the technologies, major causes of low acceptance among rural households in India, and strategy to enhance adoptability of the technology.

2. CLASSIFICATION OF BIOMASS COOKSTOVE

The biomass cookstove classification is most important to meet user choices according to locally available resources. A single cookstove cannot perform similarly when different biomass fuels are used for cooking, because of its specific design for fuel properties. For instance, a cookstove is design for deshi babool (*Acacia nilotica*) wood and it operates with other fuel say groundnut (*Arachishypogaea*) shell pellet, the energy output would differ as to the energy value of fuel input changes [10]. The woody biomass, agricultural residues, charcoal, dung cake, leafy biomass, sawdust etc. are used as a fuel for a biomass cookstove [12,15,16]. Therefore, improved cookstoves are separately designed as per the locally available biomasses and energy needs of a family [10]. Moreover, the development of multi-fuel cookstoves should be stressed because of biomass diversity in local areas rather than particular biomass cookstove. In concern with the stove materials, a cookstove may be made of mud, ceramic, cement, and other locally available materials which defined the durability of stoves. The users may opt a stove as per their convenient which either be a fixed type or portable cookstove. The benefit of a portable stove over a fixed type is that it could be carried anywhere. The fixed types of stoves used in the houses are either made of mud or cement which are relatively heavier in weight. These stoves could be fitted with a chimney to vent-out the smoke [15]. Whereas, the advancement in technology and upgraded the naturally draft stove into a forced draft stove. The benefit of the forced draft stove is that it provides sufficient amount of air and thus improves the combustion efficiency resulting in higher thermal efficiency and lower emissions of household air pollutants. The air is supply to the combustion zone through a small blower/fan [10]. Nowadays, the cookstoves are also being used as a multi-function device, i.e. for cooking, operating a small blower, LED lighting, mobile charging etc. by tapping the waste heat of the stove to produce DC power using a thermoelectric generator [17]. Thus, the biomass cookstove is classified based on a) fuel used for cooking, b) materials of construction, c) portability, d) function of the cookstove, e) types of application, f) draft use, g) based on a chimney to vent-out gaseous pollutants and particulate matter, h) types of combustion, and i) technology used [15]. The details of cookstove classifications are given in Fig. 10.2.

3. RECENT ADVANCEMENT IN COOKSTOVE DEVELOPMENT

Raman et al., [3] designed, developed and evaluated the performance of a multi-fuel forced draft stove coupled with a thermoelectric generator (TEG) as shown in Fig. 10.3. A forced draft stove incorporates a small blower to provide sufficient amount of air to improve combustion efficiency. The blower operation requires continuous power by means of a battery. The frequent power-cut in the rural areas is a major problem and could not make it possible to charge the battery all time. The refore, it becomes an unviable cooking solution. The coupling of TEG enables to use of waste and some heat from the combustion chamber to generate DC power and thus supply an uninterrupted power to operate the blower.

A TEG mainly comprises of a TEG module, a heat receiver, a heat sink, and a small blower. It works on the principle of See back effect where one side of TEG is acted as a hot junction and another side as the cold junction. The difference in temperature between hot and cold junction generates an electromotive force and thus the DC power. They reported about 44% overall efficiency of the stove along with 4.5 watt DC power generation at a temperature difference (ΔT) of 240°C between the hot and cold side. Around 0.83 watt power was consumed by the blower and 3.67 watt was available for mobile phone charging and LED lighting. Meanwhile, the higher cost of around Rs. 3,500 further makes it unviable for rural households. Carbone et al., [18] devised a novel whirl cookstove by changing the air supply pattern into the combustion zone in natural draft mode. The air into the combustion chamber was supplied tangentially through the lateral slots of the cookstove. The whirl modification creates a better air movement in which the flame propagated in a helical motion in upward direction resulted in better air fuel mixing, enhanced residence time of burning, more uniform combustion, and thus reduced pollutant emissions. The stove was fitted with two

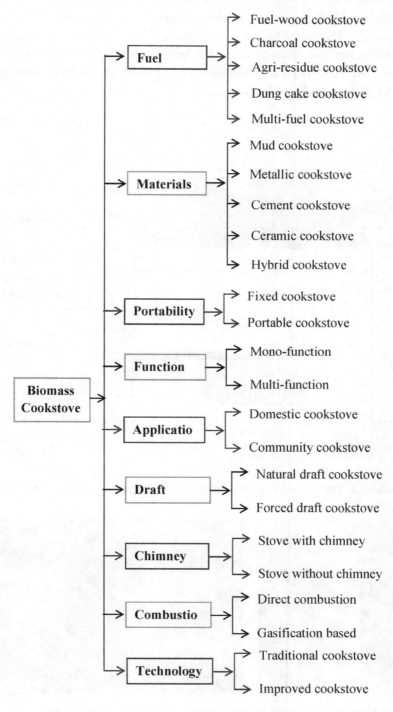

Figure10.2: Classification of Biomass Cookstove [15]

diametrically opposed chutes for biomass feeding as shown in Fig. 10.4. The thermodynamic efficiency of the stove was found to be around 68% with reduction in particulate matter emissions by 60%. The better efficiency and lower emissions enable to use the stove in rural household. The capital cost of the cookstove was estimated to be around Rs. 2250/-.

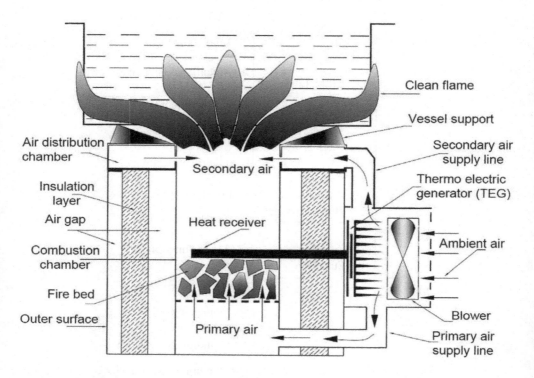

Figure10.3: TEG Based Forced Draft Stove Developed by Raman et al., [3]

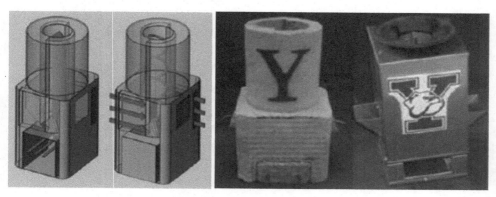

Figure10.4: Whirl Air-Supply Based Improved Cookstove [18]

Sakthivadivel and Iniyan [19] designed a fixed bed advanced micro-gasifier stove and tested its performance with tamarind seed pellet. The stove was designed by considering energy requirements of a four member's family, one meal per person per day, cooking time of 15 minutes, and specific gasification rate (SGR) of 30 kg m^{-2} h^{-1}. The combustion chamber was 11 cm in diameter and 15.5 cm in height. To supply the secondary air for burning of producer gas, the holes were skewed at 45° as depicted in Fig. 10.5 to achieve a better turbulence. They achieved a thermal efficiency of around 35% and 17.3% as exergy efficiency. Even the flame temperature was recorded to be higher (800-1000°C) than the conventional cookstove (700-800°C). Kumar and Panwar [10] designed a novel gasifier biomass stove which operates in dual draft mode efficiently. The problem of interrupted power supply to charge the battery of the blower has solved by designing the stove in such a way that it could operate in natural draft mode. If the battery is charged, it could operate in forced draft mode.

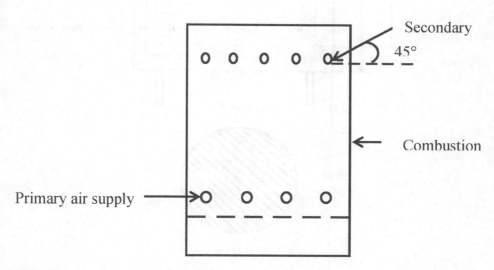

Figure10.5: Schematic of Combustion Chamber of Advance Micro Gasifier Stove [19]

The cross-sectional view of the cookstove is shown in Fig. 10.6.

The thermal efficiency of around 33.44% in natural draft mode and 36.56-36.79% in forced draft mode with multi-fuel option enable the users to adopt the cookstove in their daily lives.

The lower household air pollution and cheaper price of about Rs. 770/- with novel features attract the stove utmost. Gupta et al., [14] designed a natural draft portable multi-fuel stove with two reactors in one single assembly, which enables the users to feed the fuel either from the top or side-bottom of the cookstove. The

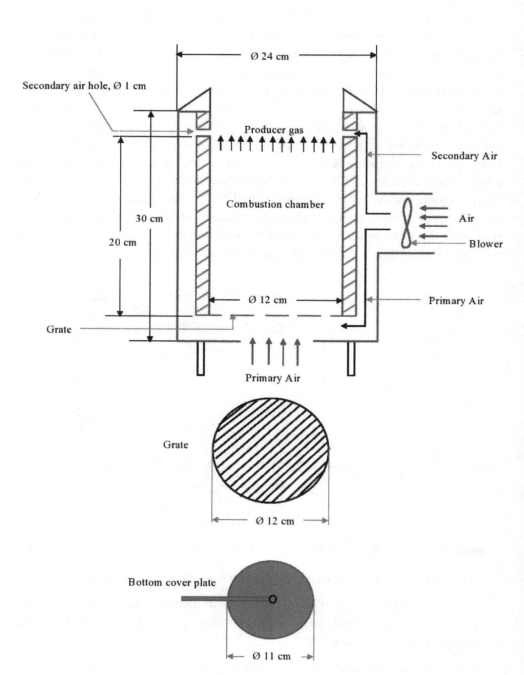

Figure 10.6: Cross-Sectional View of Dual-Draft Novel Stove [10]

performance of the stove was evaluated with dung cake, crop residue, wood, coal, and charcoal. The thermal efficiency of the stove was found to be around 27.31%, 24.58%, 29.45%, 34.72%, and 32.59% with the above respective fuel. However, the thermal efficiency was found to be lower with crop residues than the minimum efficiency of a natural draft stove suggested by the Bureau of Indian Standard BIS: IS-13152 (Part-I). The stove is depicted in Fig. 10.7.

Figure 10.7: Natural Draft Multi-Fuel Improved Cookstove [14]

4. POTENTIAL TO MITIGATE GHG EMISSIONS THROUGH ICS

The burning of solid biomass for cooking and heating contributes about 25% of the total black carbon (BC) emission globally [20]. The BC is a carbonaceous aerosol particular matter released due to incomplete combustion of solid biomass, conventional and other carbonaceous fuel [13] which has 680 times higher global warming potential than carbon dioxide [6]. It has reported that the mitigation of BC could be the second-largest modifier of climate change after carbon dioxide [13,20]. Thus, it becomes important to reduce the emissions of carbon dioxide and BC as well by adopting a clean cooking mechanism. The incursion of ICS could play a vital role in reducing the emission of biomass burning. Therefore, this section presents the potential of ICS in emission reduction in the term of carbon dioxide equivalent (CO_{2e}). The GHG reduction potential of some ICS is presented in Table 10.1. A twin-mode improved biomass cookstove developed by Kumar and Panwar [10] is efficient to minimize CO_{2e} emissions by 6.65-7.04 tons than the traditional stove annually. The novel feature of the cookstove allows the user to use in natural draft mode as well as in forced draft mode efficiently. The producer gas cookstove developed by Panwar and Rathore [21] for community cooking enables to save a significant amount of around 7.16 tons of carbon dioxide emission annually. Whereas, Dissanayake et al., [22] reported a simple improved cookstove made of cement called "Mirt" in Ethiopia is able to save around 0.94 tons CO_2 per year by saving 25% of the fuelwood annually. The Ethiopian government has set a target to distribute 11.45 million units of the stove in the country and thus to be able to mitigate about 10.77 million tones CO_2 yearly.

Table 10.1: GHG Reduction Potential of Improved Biomass Cookstove

Cookstove type	Working fuel	Mitigation potential of GHG (CO_2 equivalent)	Reference
Gasifier based stove (2.95-3.15 kW)	Babool wood and ground nut shell pellet	6.65-7.04 tons year^{-1}	[10]
Producer gas stove (5 kW)	Babool wood	7.16 tons year^{-1}	[21]
Natural draft simple cookstove	NA	0.94 tons year^{-1}	[22]

NA-Not available

Thus, it is evident that the burning of solid biomasses in improved cookstove saves significant amount of GHG emission.

5. POLICY FRAMEWORK FOR DISSEMINATION OF IMPROVED COOKSTOVE IN INDIA

The first National Program for Improved Cookstove (NPIC) was launched in the year of 1984 by the Ministry of Non-Conventional Energy Sources (MNES) now Ministry of New and Renewable Energy (MNRE), Govt. of India with the ambitious goals of a) reducing the burden on local forest by utilization of biomass in an efficient burning device, b) improving the household health, and c) reducing the drudgery of women through time-saving in fuel collection. The target was to disseminate of 120 million improved biomass cookstove in twenty-three states of India. Under this program, the government provided subsidy of 50% to the cookstove manufacturers. However, by the end year 2002, the program was closed with a meager achievement of around 28% of the target only. The major failure of this program was reported to be the transfer of subsidy to the manufacturer rather than consumers resulted in ignorance of user's preference in cookstove designing [12]. Another program of "National Biomass Cookstove Initiative (NBCI)" was launched in the year of 2009 by MNRE, Govt. of India to foster design and manufacturing of advanced generation cookstove with higher efficiency and low emissions. The primary aim of this program was to deploy the cookstove in those households who were dependent on a traditional stove [11]. The program countered the previous short-comes of NPIC by setting up of cookstove testing centres, certification and monitoring facilities, and advanced research and development laboratory. The program also incorporated the revised testing protocols of cookstove with minimum thermal efficiency of 25% for natural draft and 35% for forced draft cookstove. Four new cookstove testing centers were established to strengthen the program. These testing centers facilitate the testing of cookstove performance as per the Bureau of Indian standard (BIS- IS: 13152; Part-I) and provide certification to sale those cookstoves in the Indian market [23]. In the year of 2014, third program name "Unnat Chulha Abhiyan" was launched with an aim to disseminate 2.75 million units of improved cookstove by March 2017; however, the scheme failed to achieve the set target and met only1.3% of the target [9].

6. PROBLEMS OF LOWER ACCEPTANCE OF IMPROVED BIOMASS STOVE

The ignorance of user preference during the design of cookstove by the manufacturer is a major cause of lower acceptance of cookstove. Even the low-quality standard of the cookstove is reported to be another problem [24, 25]. The newly designed stoves are expensive which cost around Rs. 1,500 to 6,500 and users in India are unable to buy because the monthly income of around 75% of rural households is Rs. 5,000 only [10, 26]. Therefore, it is necessary to designed and developed stoves which do not meet technical efficiency only but to have social, cultural and economic viability too [12]. Moreover, the lack of institutional infrastructures like research and development centers, training facilities, technologies and innovation exchange center, support and services are the basic need to fascinate the newly evolved technologies, however, these supports are limited in the country. The poor adaptability of improved cookstove in households is also due to the unfavorable government policy. During the initial stage of the National Program of Improved Cookstove, the government used to pay the subsidy to the cookstove manufacturers. As a result, the manufacturers never paid attention towards minimizing the cost of cookstove and ease in design according to users' choices [15]. Many of developed cookstoves do not operate with any size of fuelwood and required changes in size and shape; this is another burden for the users to use locally available biomass in the improved stove [27].

7. STRATEGY TO ENHANCE ADOPTABILITY OF ICS– AUTHOR'S VIEW

The most of the cookstove sold in India so far having thermal efficiency between 25 to 30% only. The current scenario required a rigorous effort to develop stove having thermal efficiency greater than 45% by keeping the users' preferences in mind while designing. This could be done by establishing state-of-art laboratories for research and development through collaborative work between the government and private sector. The diverse biomass availability in local areas limited to use a specific design of cookstove and hence more attention required to develop multi-fuel stove with better efficiency and emissions. Moreover, the qualities of biomass available locally throughout the year change. To encourage the "advance fuel-supply-chain system like LPG" through the collection of locally available biomass and making pellets could be another solution towards clean cooking which offer 30-50% higher energy efficiency than the solid biomass. However, this technology seems to be costly. The government should stress to deploy such technologies in Block and Panchayat level and encourage the entrepreneurs' to start-up a business with manufacturing of clean fuel through better subsidy opportunity. More cookstove testing centres would be required to facilitate performance certification in time. Most of the cookstove performed well in the laboratory but failed in actual use, thus, its actual performance in the field must be done before approval. The ministry should mandate in policy to enforce the centres to act as a think tank in technology improvements and provide complete scientific and technological support to the manufacturers in public-private partnership mode. The tested stove must be labeled with thermal efficiency, emissions rating, working fuel, and fuel-saving capability etc. for ease of selection by the users. Another important parameter is that the manufacturer should provide warranty and repair service facilities to the users to enhance the trust in technology. The higher cost of cookstoves is another major problem. The purchasing ability of the customers could be enhanced by providing better subsidy and even through loan basis.

8. CONCLUSIONS

To mitigate household air pollutions, the use of locally available biomass through ICS could be the best solution. The ICS is able to curb the household pollutants and GHG emissions significantly. However, the current policy and programs are failed to disseminate those technologies among rural households. The scanty achievement of 28% out of 120 million ICS in the first program (NPIC) and 1.3% of 2.7 million stoves in "Unnat Chulha Abhiyan" clearly indicates the lack in policy and trust of technologies among the rural households. The scientific community and policymaker should incorporate the preference of users' requirement before designing a biomass cookstove. The multi-fuel cookstoves are now emerging in the market which would enable the users not to stick with specific biomass but use many varieties of locally available biomasses. The higher cookstove price is found to be the major cause of lower acceptance and thus reducing the price of cookstove would be a game-changer. It has also been reported that only 14% of the rural households are aware of the technology, and thus, the awareness program/demonstration through the testing centres, institute, NGOs, and private agencies could enable to understand the benefits of ICS among rural residents.

REFERENCES

1. Abrar MM. Power cut off and power blackout in India a major threat: an overview. International Journal of Advancements in Research & Technology. 2016; 5(7): 8–15.

2. Hiloidhari M, Das D, Baruah DC. Bioenergy potential from crop residue biomass in India. Renewable and Sustainable Energy Reviews. 2014; 32: 504-12.doi.org /10.1016 /j.rser.2014.01.025.

3. Raman P, Ram NK, Gupta R. Development, design and performance analysis of a forced draft clean combustion cookstove powered by a thermo electric generator with multi-utility options. Energy. 2014; 69: 813-25. doi.org/10.1016/j.energy.2014.03.077.

4. Waris WS, Antahal PC. Fuelwood scarcity, poverty and women: some perspectives. IOSR-JHSS.2014; 19: 21-33.

5. Mehetre SA, Panwar NL, Sharma D, Kumar H. Improved biomass cookstoves for sustainable development: a review. Renewable and Sustainable Energy Reviews. 2017; 73:672–687. doi.org/10.1016/j.rser.2017.01.150

6. Pratiti R, Vadala D, Kalynych Z, Sud P. Health effects of household air pollution related to biomass cook stoves in resource limited countries and its mitigation by improved cookstoves. Environmental Research. 2020:109574. doi.org /10.1016/j. envres.2020.109574.

7. Menghwani V, Zerriffi H, Dwivedi P, Marshall JD, Grieshop A, Bailis R. Determinants of cookstoves and fuel choice among rural households in India. Eco Health. 2019;16 (1) :21-60.doi.org/10.1007/s10393-018-1389-3.

8. Kumar H. Design, Development, and Performance evaluation of a Wood Gas Cookstove with Thermoelectric Power Generation System (Master dissertation, MPUAT, Udaipur). 2015.https://krishikosh.egranth.ac.in/handle/1/5810023128.

9. Patnaik S, Tripathi S, Jain A. A roadmap for access to clean cooking energy in India. Asian Journal of Public Affairs. 2019; 11(1): 1-76.

10. Kumar H, Panwar NL. Experimental investigation on energy-efficient twin-mode biomass improved cookstove. SN Applied Sciences. 2019; 1(7):760. doi.org /10.1007/ s42452-019-0804-x.

11. Venkataraman C, Sagar AD, Habib G, Lam N, Smith KR. The Indian national initiative for advanced biomass cookstoves: the benefits of clean combustion. Energy for Sustainable Development. 2010; 14(2):63-72. doi:10.1016/j.esd.2010.04.005

12. Lambe F, Atteridge A. Putting the cook before the stove: a user-centred approach to understanding household energy decision-making. Stockholm Environment Institute. 2012.

13. Ravindra K. Emission of black carbon from rural households kitchens and assessment of lifetime excess cancer risk in villages of North India. Environment international. 2019; 122:201-12. doi.org/10.1016/j.envint.2018.11.008

14. Gupta A, Mulukutla AN, Gautam S, TaneKhan W, Waghmare SS, Labhasetwar NK. Development of a practical evaluation approach of a typical biomass cookstove. Environmental Technology and Innovation. 2020;17:100613. doi.org /10.1016 /j.eti. 2020.100613

15. Kshirsagar MP, Kalamkar VR. A comprehensive review on biomass cookstoves and a systematic approach for modern cookstove design. Renewable and Sustainable Energy Reviews. 2014;30:580-603.doi.org/10.1016/j.rser.2013.10.039.

16. Yevich R, Logan JA. An assessment of biofuel use and burning of agricultural waste in the developing world. Global Biogeochemical Cycles. 2003; 17(4).doi.org /10.1029 /2002GB001952.

17. Panwar NL, Kumar H. Waste heat recovery from improved cookstove through thermoelectric generator. International Journal of Ambient Energy. 2019;14:1-5.doi .org/10.1080/01430750.2019.1653978.

18. Carbone F, Carlson EL, Baroni D, Gomez A. The whirl cookstove: a novel development for clean biomass burning. Combustion Science and Technology. 2016; 188(4-5):594-610. doi.org/10.1080/00102202.2016.1139364.

19. Sakthivadivel D, Iniyan S. Experimental design and 4E (energy, exergy, emission, and economical) analysis of a fixed bed advanced micro gasifier stove. Environmental Progress and Sustainable Energy. 2018; 37(6): 2139-47. DOI 10.1002/ep.12882.

20. Serrano-Medrano M, García-Bustamante C, Berrueta VM, Martínez-Bravo R, Ruiz-García VM, Ghilardi A, Masera O. Promoting LPG, clean wood burning cookstoves or both? Climate change mitigation implications of integrated household energy transition scenarios in rural Mexico. Environmental Research Letters. 2018; 13(11): 115004.DOI: 10.1088/1748-9326/aad5b8.

21. Panwar NL, Rathore NS. Design and performance evaluation of a 5 kW producer gas stove. Biomass and Bioenergy. 2008; 32(12): 1349-52. Doi :10.1016/j.biombioe. 2008. 04.007.

22. Dissanayake ST, Beyene AD, Bluffstone R, Gebreegziabher Z, Kiggundu G, Kooser SH, Martinsson P, Mekonnen A, Toman M. Improved Biomass Cook Stoves for Climate Change Mitigation? Evidence of Preferences, Willingness to pay, and Carbon Savings. The World Bank; 2018.

23. National Biomass cookstove program, Ministry of New and Renewable Energy. http://164.100.94.214/national-biomass-cookstoves-programme.

24. Hanbar RD, Karve P. National Programme on Improved Chulha (NPIC) of the Government of India: An Overview. Energy for Sustainable Development. 2002; 6(2): 49-55.

25. Kishore VV, Ramana PV. Improved cookstoves in rural India: how improved are they? A critique of the perceived benefits from the National Programme on Improved Chulhas. Energy. 2002; 27(1): 47-63.doi.org/10.1016/S0360-5442 (01)00056-1.

26. Shrimali G, Slaski X, Thurber Mark C, Hisham Zerriffi. Improved stoves in India: a study of sustainable business models. Energy Policy. 2011; 39 (12):7543–56.doi.org/ 10.1016/j.enpol.2011.07.031.

27. Mukhopadhyay R, Sambandam S, Pillarisetti A, Jack D, Mukhopadhyay K, Balakrishnan K, Vaswani M, Bates MN, Kinney P, Arora N, Smith K. Cooking practices, air quality, and the acceptability of advanced cookstoves in Haryana, India: an exploratory study to inform large-scale interventions. Global Health Action. 2012; 5(1):19016.DOI: 10.3402/gha.v5i0.19016.

CHAPTER - 11

PRACTICAL EVALUATION APPROACH OF A TYPICAL BIOMASS COOKSTOVE

Amit Ranjan Verma*, Ratnesh Tiwari, Manoj Kumar Verma and Himanshu Kumar

*Centre for Rural Development and Technology
Indian Institute of Technology Delhi, Hauz Khas, New Delhi-110016, India
Corresponding Author

The chapter deals with performance evaluation of biomass cookstoves and identifies the main criticalities related to the testing and assessment of biomass cookstoves and provides indications for a better testing approach. Biomass has been a critical resource for the human being in meeting the fuel requirement for cooking and space heating since ancient time. Its performance evaluation can play a crucial role by providing the required information to the decision-makers. It can also help in the selection of the appropriate technology as per the requirement. Globally used test method known as Water Boiling Test (WBT) is a simulation of a cooking cycle conducted under controlled laboratory conditions. However, WBT done in the lab and field were not in good agreement. Laboratory testings are considered to be reproducible and useful to compare the performance of different cookstoves. Overall we can say that the assessment of cookstove performance in the lab can provide some valuable sets of data to identify the impact range of independent variables on cookstove performance. Field measurements are considered to give representative and reliable results. However, it requires comparatively more human resources and costsand also very time-consuming. The field tests are performed in actual use conditions demonstrates how biomass cookstoves perform with local cooks, cuisines, practices, fuel types and environmental conditions, etc. leading to higher variability in results. Field testing is the better way to verify the performance results of biomass cookstoves like reduced fuel consumption and emissions and better energy output.

1. INTRODUCTION

In India, as reported by India Census 2011, a significant proportion (around 85%) of the rural population still uses traditional biomass fuels for meeting energy needs. These fuels are wood, agriculture residues, animal dung, etc. [1]. They are burnt in traditional cookstoves. The incomplete combustion of these fuels in the traditional cookstoves leads to the emission of harmful gases like carbon monoxide and particulate matter. These emissions cause indoor pollution, which is very detrimental to health mainly for women and children who are exposed for a longer duration [2]. As per the estimation of WHO, indoor air pollution accounted for 4.3 million deaths in 2012. To be more precise, these deaths were caused due to indoor air pollution generated from cooking over coal, wood, and biomass stoves [3]. The traditional use of biomass is also expected to result in deforestation, although no literature confirms a direct link between deforestation and biomass use.

Improved Cook Stoves (ICS) presents itself as a potential technological alternative to the traditional cookstove addressing the concerns mentioned above. An ICS is a cookstove designed using certain scientific principles, to aid better combustion and heat transfer, also improving emissions and efficiency performance as compared to traditional stoves [4]. Although ICSs offer various advantages over theirconventional counterpart, there are specific issues they are yet to be addressed for the success of ICSs. These issues are stove stacking (combined use of traditional and ICS for different tasks) [5], competing applications of biomass with other activities (lighting, heating, brick-making), nutritional habits [6] and lack of agreement over methodologies for performance evaluation of stoves [7]. The present study deals with the last aspect mentioned, *i.e.* methodologies for performance evaluation. Performance evaluation is a crucial aspect to be dealt with for the success of ICSs. It can play a crucial role by providing the required information to the decision-makers and can also help in a selection of the appropriate technology as per the requirement [8]. A well-documented data on the performance evaluation of ICSs can also serve as a significant input for global climate prediction models [9].

2. NEED FOR COOKSTOVE TESTING

The information that most of the cookstove dissemination program required is obtained from various laboratory and field tests. To generate accurate, reliable, and repeatable performance data, it is essential that the cookstove test method being used is highly reliable. A standardized cookstove testing provides a quick means of results of newly designed and developed cookstoves, and it can be useful in finding benefits due to the intervention of improved cookstoves. It is imperative to develop an internationally accepted testing method for evaluation of cookstove performance as can be seen as successful in case of China cookstove dissemination programs and failed in India during the National Programme on Improved Cookstoves (NPIC, 1985). As per reports by Kishore and Ramana (2002), NPIC was a failure in India because the low performance of cookstoves along with lousy execution of the programme [10]. Due to the unavailability of an internationally accepted and acknowledged testing protocol, various testing protocols based on different approaches had been made by different research groups. However, testing protocols have been questioned due to uncertainty attached to methodologies adopted. Scientists in their works of literature have put forth the urgent requirement for making the protocols repeatable and operate close to field conditions.

2.1 Types of Testing

The performance of biomass cookstoves can be evaluated in laboratory and field-based tests.

2.1.1 Laboratory Testing

Laboratory testing is useful for design optimization of the newly designed cookstove and gives insights to developers to assess changes in performance due to different designs and features. As these tests are conducted in a controlled laboratory setting, no variability due to the external condition is observed. Laboratory test result gives parameters like efficiency, specific consumption, CO emissions, PM emissions [11]. As the real users operate in different conditions compared to laboratory testing, therefore, it is unsuitable for evaluating the real-world performance outcomes by laboratory tests [40].

2.1.2 Field Testing

In a layman's language, testing which involves actual users in their particular environment is called field testing. Studies of user outlooks toward a stove, and studies of consumers' requirements and inclinations are common market factors, that might provide hints about the method of using stove by the user, they should ideally be regarded as field tests. Some other kinds of field testing methods collect data on a stove in actual use conditions, and by their nature, unlike labs, the real conditions cannot be tight. As laboratory test does not necessarily simulate practical use scenarios for a stove, it cannot be considered as an ideal method for performance evaluation in a real situation whereas field testing can be regarded as the appropriate method in the assessment of cookstove performance. Field testing can also be used for the identification of key user constraints affecting the use and acceptance of stoves.

2.2 Performance Parameters

Performance of biomass stoves depends on various operational constraints such as characteristics of the fuel used (type, size, and moisture), sizes and kinds of pots used, fuel feeding rate, ambient conditions, ventilation levels, etc. Table 11.1 presents a summary of the parameters used for the evaluation of cookstoves and other important dependant factors [12].

Table 11.1: Parameters Used for Evaluation of Cookstoves and Other Important Dependant Factors [12]

Parameter	Function	Other important dependant factors
Efficiency	Combustion	Level of CO_2, CO, & O_2 in flue gas
	Heat transfer	Flame temperature
	Heat absorption by the cookstove materials	Mass and thermal property of the materials used in the cook stove.
	Heat loss through the flue gas	Flue gas temperature
Gaseous emission	Partial combustion of fuelwood	A suitable supply of air for efficient combustion
Particulate emission	Particulate matters carried away through flue gas (PM_{10}, $PM_{2.5}$)	Air supply level and flue gas escaping velocity

2.2.1 Thermal Performance Parameters

Thermal performance of a cookstove is characterized in terms of the following parameters:

2.2.1.1 *Power*

The regular rate of energy outduring the entire burning duration is termed as fire power by [13, 14]. Baldwin contemplates fuel and calorific value in terms of as-received basis, whereas Yuntenwi considers the 'equivalent dry fuel consumption' with assumptions on moisture content and calorific value of char. This different approach was taken up by the two researchers' results in significantly different calculated efficiency values for the same test results.

2.2.1.2 *Thermal Efficiency and Specific Fuel Consumption*

Bhatt(1982) defined specific per day consumption (SDC) and specific task consumption (STC) of fuel as the grams of wood equivalent consumed per kg of water boiled [15]. Baldwin (1987) used the term specific consumption (SC) for the same definition [16]. Despite being closely associated with efficiency, the term specific consumption has its special importance. In some instances, the cookstove might probably achieve high efficiency, however, with more significant evaporation losses from the pot, it may lead to a flawed indicator of cookstove performance. Therefore, Jetterand Kariher (2009) suggested that efficiency can be a misleading indicator of the same [17].

2.2.1.3 *Turn-Down Ratio*

The ratio between the maximum and minimum power of the cookstove is defined as a turn-down ratio, which is again an essential parameter for evaluating the performance of a cookstove [18]. According to the tests conducted by Still et al. (2011) [19] on 18 different stoves, the turn-down ratios of all stoves was found to vary from 2 to 4. 14 out of the 18 stoves tested used solid biomass.

2.2.2 Emission Performance Parameters

The emission parameters used frequently for characterization of the stove are as follows:

Emission Factor (g/kJ): This is the amount of a pollutant emitted from acookstove in grams per kg of the fuel burnt.

Indoor Concentration (mg/m^3): This is estimated as pollutant concentration in an indoor kitchen.

2.2.3 Effect of Operating Parameters

2.2.3.1 *Fuel Type*

According to Bussmann (1988) [20] and Bussmann et al., 1983 [18], the mass-loss rate of the fuel and the maximum volatiles power for open fires both depend strongly on wood species. Different types of biomass fuels have different stoichiometric air requirement and hence the combustion characteristics, resulting in variation in pollutant emissions.

2.2.3.2 Fuel Size

According to Dirks (1991), a decrease in the power output was observed with an increase in volume/surface area of fuel (hence the size of the fuel) [21]. As reported by Bhattacharya et al., (2002) the change in fuel size was found to have no significant effect on the efficiency of the cookstove. Although it is evident that increases in fuel size led to a slight increase in CO emissions and a decrease in NO_x emissions [22].

2.2.3.3 Fuel Moisture Content

Bussmann (1988) found that as moisture content raised from 0-25%, the maximum volatile power initially decreases to about 5% moisture content, and after that, it becomes constant at design power [20]. Claus et al. (1983) tested TungkuLown stove and reported that changing the moisture content from 0-6% using small sized fuel caused no significant change in efficiency and combustion performance [23].

2.2.3.4 Pot Size and Lid

According to the reports given by Bhattacharya et al. (2002), pot size was found not to affect the cookstove efficiency [22]. In contrast, Mukunda et al. (2010), in their work, have reported that pots with large diameter to height ratio resulted in increased heat transfer area for a given volume of the container [24].

De Lepeleire and Christiaens (1983) observed reduced evaporation losses on using pot lids at high power near boiling temperatures [23].

3. STOVE TESTING PROTOCOLS

3.1 History and Evolution of Testing Protocols

ICS testing protocols came into existence in the 1980's. The Intermediate Technology Development Group (ITDG), now known as Practical Action, first defined a procedure for testing of cookstoves in laboratory and field [25]. A few years later, between 1982 and 1985, three different protocols, namely Water Boiling Test (WBT), Controlled Cooking Test (CCT), and Kitchen Performance Test (KPT), addressing different testing needs were developed. These protocols were developed by Volunteers in Technical Assistance (VITA) based upon ideas from ITDG and Eindhoven Woodburning Stove Group [26]. "Biomass stoves: Engineering design, development, and dissemination" [13] by Dr. Samuel Baldwin presents the first revision and discussion on VITA's WBT. Since then, it has been the most widely-cited reference for researchers engaged in stove development (referred to as WBT 2.0). The popularity of Baldwin and Vita's publications led to the widespread adoption of water boiling tests. The NPIC (National Programme on Improved Chulhas) and the CNISP (Chinese National Improved Stoves Programme) were the first two big-scale dissemination programs of ICSs launched in India and China during 80's (1982-1983) [4, 27]. The two Countries created specificmethods for testing cookstoves, which were formalized and published after a few years. The first version of the Indian Standard on Solid Biomass Chulha-Specification (acronym BIS) is dated 1991, which was revised in 2013. It has been cited by many authors and is available only in the form of a draft [28]. The first version of the Chinese Standard that can be found in the literature back to (Chinese Standard, 2008), but testing methodologies and benchmarks were established since the launch of the program. In 2003, under the Shell Household Energy and Health project, Dr. Kirk Smith and Rob Bailis, in collaboration with researchers from the Aprovecho Research Center, performed a revision of the VITA's protocol. The project was commissioned to the University of California-Berkeley. WBT version 3.0 [29] and CCT, KPT versions 2.0, were developed between 2003 and 2007. A phase named Cold-Start marked one of the most significant variations in the WBT. The group Engineers in Technical and Humanitarian Opportunities of Service (ETHOS), initiated a technical committee, led by Dr. Tami Bond of the University of Illinois along with Partnership for Clean Indoor Air (PCIA). Further, revised the WBT to version 4.1.2 (2009), including instructions for emissions measurement and testing of non-woody solid, liquid, or gaseous fuels. A protocol named Emission & Performance Test Protocol (EPTP) was developed by Colorado State University and Shell Foundation, in collaboration with cookstoves manufacturers Philips and Envirofit in 2009 [23]. This protocol was developed based on the updated WBT but aimed at the optimization of reproducibility.

In 2010, two more protocols named Adapted Water Boiling Test (AWBT) [29] and Heterogeneous Testing Procedure (HTP) were developed by researchers from GERES Cambodia and the SeTAR Centre based in Johannesburg, respectively. The former was developed as a modification of WBT 3.0 and focused on following standard cooking practices in the target area, instead of relying on standardized parameters. The later protocol was developed for the requirements of a GIZ project (named Pro-BEC) on domestic stoves. The procedure concept, equipment, and calculated parameters of this protocol were quite different from that of WBT. In 2012, a workshop named ISO-IWA, with a participation of more than 90 stakeholders from 23 countries, was scheduled in The Hague (IWA, 2012) [30]. The workshop was organized by the Global Alliance for Clean Cookstoves and provided interim guidance for rating cookstoves on four performance indicators: efficiency, total emissions, indoor emissions, and safety; for each indicator, multiple Tiers of Performance (0 to 4) were defined, to set a hierarchy in the ICSs technological advancement. The current version of WBT 4.2.3 (2014) includes results from the ISO- IWA workshop and tiers of performance, as well as indications coming from other research groups. Figure 11.1 presents the chronology of protocols development.

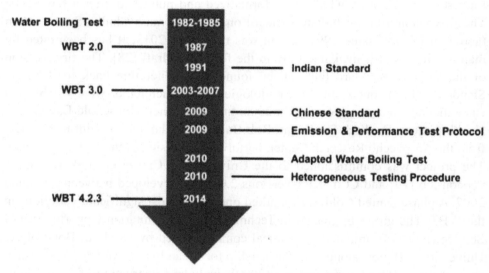

Figure 11.1: Chronology of Protocols Development [11]

Tables 11.2 and 11.3 present the summary of different laboratory and field protocols, respectively.

3.2 Performance Evaluation of Biomass Cookstoves

As the present work involves the testing and evaluation of biomass cookstoves in the lab as well as in the field, it is necessary to study the literature reporting experimental work carried out by some researchers. Ezzati et al. (2000) proposed a comparative study on the emissions and residential exposure from conventional and upgraded modern cookstoves in Kenya [3]. This study showed that ceramic wood burning stoves reduced daily average suspended particulate matter

Table 11.2: Summary of Some Cookstove Laboratory Testing Protocols

Particulars	WBT 4.2.3 (2014)	Indian BIS (2013)	AWBT (2010)	HTP (2010)	EPTP (2009)	Chinese (2008)
Preparation and procedure for testing						
Number of phases	3 (2 high power 1 low power)	1 (high power)	2 (1 high power low power)	3 (1 high power 1 low power)	3 (2 high power 1 low power)	2 (1 high power 1 low power)
Minimum no of pots	1	2	1	3	1	1
Pot size	7 l	Depend on firepower of stove.	Ones most commonly used by the local stove end users.	6.4 l or 3	A table provided to decide the quantity of water depending on firepower of the stove and initial water temperature.	Depends on fire powder of stove.
Quantity of water in pot	5 l in large pot and 2.5 l in small pot.	Depend on firepower of stove.	3 -5 l, or enough water to fill 2/3 of the cooking pot used for the test.	80% of the pots volume	4-6 l	Depends on fire power of stove 5 kg to 9 kg.
Initial temperature of water	Ambient temperature	23 ± 5°C	Not specified	Not specified	4-30°C	Not specified
Maximum temperature of water	Boiling temperature	95°C	Boiling temperature	80°C	90°C	Boiling temperature
Use of pot lid	No	Yes	No	Yes	Foam insulation (High power only)	Yes (High power only)

[Table Contd.

Contd. Table]

Contd. Table]

Particulars	WBT 4.2.3 (2014)	Indian BIS (2013)	AWBT (2010)	HTP (2010)	EPTP (2009)	Chinese (2008)
Fuel dimension	1.5 × 1.5 cm^2	3 × 3 cm^2 for family type 4 × 4 cm^2 for community type	Not specified	Not specified	1.5 × 1.5 cm^2	Not specified
Length of fuel pieces	Not specified	Half diameter/length of combustion chamber	Not specified	Not specified	Not specified	Not specified
Moisture content of fuel	6.5% or 10%	5±1 %	15% (for fuelwood) 5%	Not specified	4-10%	Not specified
Fuel feeding rate	Depending on the type of stove, fuel feeding may be batch or continuous type, control the fire with.	Continuous fuel us 1 feeding an at interval of 6 min. The fed batch type cook stove shalve 1 be filled with the fuel as specified by the manufacturer.	Batch feeding only at start.	Batch feeding.	Stove manufacturers. specifications if provided, otherwise feed the fuel to heat the water rapidly during old and hot start phase and use minimum amount of fuel while simmering.	Not specified

[Table Contd.

[Contd. Table]

Particulars	WBT 4.2.3 (2014)	Indian BIS (2013)	AWBT (2010)	HTP (2010)	EPTP (2009)	Chinese (2008)
Ignition	Depends on local habits) start test after the fire has caught.	Use kerosene as kindling, start test as soon as flame develops	(depends on local habits) start test when kindling is exhausted	(Depends on local habits)	(Depends on manufactures indications; if not present, kindling materials are suggested based on fuel type)	Left to tester's discretion) start test {when the fuel starts to burn}
Phase duration	45 min (low phase)	60 min	Not specified	Not specified	45 min (low phase)	Not specified
End of test	After 45 minutes of simmer phase the test is over	When there is no visible flame left in the combustion chamber, the test is over.	As fuel exhausted to the point that the temperature of the water drops 3^0C below the boiling point, the test is over and the	As pot-3 reaches 80^0C emission measurement is stopped and the test is over.	After 45 minutes of simmer phase the test is over	As the temperature of water drops to 95^0C, due to insufficient fuel power, the test is over.
Emission measurement	Gases- (electrochemical and NDIR method with pros and cons) Particulate Matter- (gravimetric and optical methods with pros	Gases- NDIR Particulate matter- Gravimetric	Not specified	Gases- electrochemical and NDIR	Gases- electrochemical and NDIR Particulate matter- Gravimetric and optical	Not specified

Table 11.3: Summary of Some Cookstove Field Testing Protocols.

Particulars	CCT	KPT	UCT
Description	It measures stove performance in comparison to traditional cooking methods when a cook prepared a predetermined local meal in a controlled setting using local fuels, pots and practices.	It is used to evaluate stove performance in real-world settings. It is designed to assess actual impacts on household fuel consumption.	Similar to the CCT, but meal is not predetermined, and the cooks prepare what they want, with their own fuels and are not instructed to change their cooking practices in any way.
Application	Determined what performance is possible in households under controlled conditions, but not necessarily what is achieved by households during daily use.	KPTs are conducted in the course of an actual dissemination effort with real populations cooking normally and give the best indication of real-world performance.	Conducted to determine real-world stove performance.
Primary Metrics	Fuel use per amount of food cooked, cooking time, firepower (emission can be measure, but are not of the standard protocol)	Fuel use per household per day, fuel use per person per household per day (emissions can be measured, but are not part of the standard protocol)	Emission Factors: • g-pollutant/kg fuel • g-pollutant/MJ fuel • g-pollutant/person meal • g-pollutant/g food ○ Emission rates ○ g pollutant/min ○ Modified combustion efficiency (CO_2/ [CO_2 + CO] molar
Important notes	Results are more variable than controlled laboratory testing, which means larger sample sizes are required to obtain reasonable confidences in the outcomes. • Estimates are generally location and cuisine specific. • Can be done in homes or at local facility/meeting area.	• Results capture variability of normal stove/fuel use patterns in homes. • Required large sampled sizes. • Logistical requirements of conduction field campaign are substantial.	• Variability is relatively high requiring more samples than controlled testing to achieve similar confidence in estimates. • Must be conducted in homes of participants. • Logistical requirements of conduction field campaign are substantial, but no training of participants is required.

concentration by 48% (1822 μg/m³; 95% C.I. 663-2982) during the active burning period and by 77% (1034 μg/m³; 95% C.I. 466-1346) during the smoldering phase. Ceramic stoves also reduced the median as well as reduced 75% and 95% emission during the burning period and 95% during the shouldering phase, and therefore overall makes the emission rate lower. The highest reduction in daily emission rate was achieved, when wood is replaced by charcoal as burning agent, where average emission concentrations dropped by 87% (3035 μg/m³; 95% C.I. 2356-3500) during the burning period and by 92% (1121 μg/m³; 95% C.I. 626-1216) when during smoldering phase. These results inferred that charcoal based improved wood stoves were feasible options for reduction of indoor air pollution.

Berrueta et al. (2008) worked on the energy performance of wood-based cookstoves in Michoacán, Mexico. From the results of this work, it is a clear indication that the overall performance of the WBT was highly enhanced in rural communities [31]. In estimating the benefits of improved stoves for rural communities, field testing is a crucial part. In the CCT, for tortilla making, the main cooking task in Mexican rural households, Patsari stoves showed fuelwood savings ranging from 44% to 65% in relation to traditional open fires (n=6; before and after Patsari adoption of 67% (n=23; P<0.05) in rural households exclusively using fuelwood. Similar energy savings of 66% for fuelwood and 64% for LPG, respectively, were also observed in households using mixed fuel.

MacCarty et al. (2008) worked on the performance of cookstove. The study includes both the field and lab studies of rocket stoves. It consists of a comparative study on the open fire and traditional stoves in Tamil Nadu state, India focusing on time to cook, fuel use for cooking, total emissions of pollutant gases, and indoor air pollution [32]. According to this report, the in-field use of Rocket stoves (without pot skirts) proved to be more fuel efficient than that of traditional cookstoves and reduced fuel used from 39% to 47% compared to the Three Stone Fire. In the context of the emission reduction, the non-chimney stoves were about 45% emission savings as compared with the traditional stoves and about 50-55% in comparison to the three-stone fire. In emission reduction, the rocket stoves were more efficient than the traditional stoves as well as the three-stone fire.

Roden et al. (2009) reported the studied laboratory and field level measurement of particulate matters and carbon monoxide emissions from traditional and technologically improved cookstoves. From the above study, the particulate emission from the field level cooking was found to be three times as that of laboratory level simulate cooking [33]. The single scattering albedo (SSA) of the emissions was very low in both lab and field measurements, averaging about 0.3 for laboratory method and around 0.5 for field method, which indicates that the primary evolving particles cause climate warming. These researchers have observed that technologically designed improved cookstoves would significantly reduce PM and CO emission as that of the conventional cookstoves.

Adkins et al. (2010) studied field testing and survey evaluation of household biomass cookstoves. According to the two studies conducted in the assessment of the performance and usability of Household biomass cookstoves under field conditions in rural sub-Saharan Africa, the manufactured stoves, in general, yield a substantial reduction in specific fuelwood consumption relative to the three-stone fire. However, the result varies by type of stove and type of food cooked. Survey data suggested that the efficiency of the cookstoves depends upon the saving of the fuelwood as well as the combination of this and other factors, which includes cooking time, stove size, and ease of use [34].

Mac Carty et al. (2010) conducted a study on the performance of 50 different stove designs using Water Boiling Test (WBT) Version 3.0 and compared the fuel efficiency emission of carbon monoxide (CO) and particulate matter (PM). The results reveal that the stoves without well-designed combustion chambers were more fuel efficient compared to the three-stone fire; however, the emission of CO and PM increases to a certain extent [35].

Jetter et al. (2012) took newer cookstove technologies under consideration. They performed an extensive analysis of emissions and fuel efficiency at two moisture content levels, which were examined under laboratory level operating conditions using WBT protocol version-4 [17]. One traditional stove, 3-stone fire, was also tested under the same parameters. To develop an emissions database for cookstoves, emissions rates, and other factors are calculated based on cooking energy applied, cooking type, fuel energy, and fuel mass. Along with the emission rate; Cooking power, energy efficiency, and fuel use are also calculated under the WBT protocol. One of the technologically advanced cookstoves showed the lowest $PM_{2.5}$ emissions rate of 74 mg/MJ compared to 700-1400 mg/MJ delivered from the base-case open 3-stone cookfire. The highest thermal efficiency was found to be 53%, whereas three-stone cook fire has a thermal efficiency of 14-15%. Though the Baseline 3-stone fires had very high MCE of 96-97% but comprised of 14-15% Overall Thermal Efficiency (OTE) whereas with similar MCE, Several cookstoves (Sampada, Mayon, StoveTec, Berkeley, Envirofit) showed at least 2-fold greater OTE. Charcoal stoves showed comparatively low MCE due to high CO emissions. Although forced-draft stoves typically noted for higher MCE but it did not exhibit high MCE all the time. A natural-draft cookstove with a top-lit up-draft (TLUD) design had the highest MCE and OTE but requires processed, low-moisture, pellet fuel. The most of the cookstoves emit less CO and $PM_{2.5}$ per unit volume of water per time than the 3-stone fire base-case. A forced-draft stove (Philips fan) and charcoal stoves had notably low emissions.

Arora et al. (2014) performed a comparative study of Indian biomass cookstoves evaluating protocols and Water Boiling Test [28]. In this study, two cookstoves manufactured by Phillips, Philips Natural Draft (PN), and Philips Forced Draft (PF) were taken for assessing their performance with different protocols. It was observed that both the cookstoves gave a similar value of TE using both protocols implemented; whereas, CO emission factors were ~39% and ~47% higher in BIS test because the above method is dominated by smoldering conditions compared to WBT in PF and PN cookstoves, respectively. Particulate matter emissions were found to be ~55% lower in BIS test compared to WBT in PN cookstove, which could be due to the oven dried wood is usually used for the study in the BIS test. While the emissions of ignition phase were included, it was observed to increase the total PM and CO emissions during cold start by 45–70% when mustard stalks were used as burning material in comparison with wood chips and kerosene. When the cookstoves are evaluated using BIS Test protocol, it was found that average CO emissions were found to increase by ~68% and ~48% in PF and PN cookstoves, respectively, with 15 minutes of fuel feeding interval. The result also shows that change in combustion conditions during the two different test protocols did not affect the energy parameters; however, the effect on CO and PM emissions was significant. From this report it can be concluded that testing of cookstoves with different testing protocols infers a bulk of information, however, it creates difficulties in evaluating the actual performance of cookstoves.

Raman et al. (2014) made a study to review and analyze the existing test protocols which are used for the assessment of the performance of the cookstoves [12]. According to the analysis made by the researchers, an improved test method for performing WBT was proposed. And also found that the existing test methods have not considered the estimation of heat gained by the making material of a cookstove. From the total heat absorbed by the cookstoves, a part of the heat is available during the simmering phase, and the construction materials of the cookstove absorb the significant part of the energy. The total heat absorbed was divided into three substantial parts; heat gained by the construction material, the heat released during the simmering phase, and the heat withstand by the materials (after the simmering phase), which extensively influence the efficiency of cookstoves. For evaluation of the performance and further improvement of the cookstoves, estimation of these heat values is needed. Estimation of the recoverable heat energy absorbed by the construction material has to be needed. Therefore, the proposed method is on the estimation of this heat energy by introducing a fourth phase in WBT. To avoid the complications arising in maintaining the water at a specified temperature, in the proposed water boiling test, the temperature of the water is kept just close to the boiling point. The complete cycle WBT has been proposed for the estimation of the performance of the cookstoves along with the overall performance of the different phases of WBT as well as the turn-down ratio.

Zhang et al. (2014) proposed a study on differentiating test methods related to the stove operation and data analysis techniques, the fueling procedure, the endpoint selection, the choice of metrics, and other factors [36]. Analysis of the influences of these differences was done by using an induction heater. The result of the above comparative study showed that the use of a pot lid and the selection of the endpoint of the test have the most considerable influence on the performance of cookstoves. It infers that by differing the test methods, the same cookstove plus fuel combination can give different performance. Some metrics present in the result of popular tests should be reviewed. The different operating conditions influence a lot on the thermal efficiency, gaseous and particulate emissions of the domestic cooking stove, which will definitely affect the performance rating of any stove. A rigorous scientific review, including conceptual and mathematical, should be performed for all methods implemented for the cookstove testing, whether new, whether generally accepted, or whether enforced as legal requirements. Though some metrics are generally agreed by the "stove enthusiast" community, they are needed to be reviewed in enhanced perspectives.

3.3 Uncertainty in Testing

There are two factors that contribute to the uncertainty in stove testing: instrument uncertainty of the measurements; and the uncertainty in contribution owing to repeatability issues in measurements. Since the physical phenomena in a cookstove, as well as boiling of water, have several parameters that are uncontrollable in laboratory scale testing, the later factor attributes substantially to the overall uncertainty in the results of cookstove testing [39, 41]. L'Orange et al. (2012) conducted water boiling tests using WBT 3.0 and EPTP protocols and found that EPTP protocol reduces uncertainty in results during testing of cookstoves (Fig. 11.2) [37, 38].

4. CONCLUSIONS

This work aimed to identify the main criticalities related to the testing and evaluation of biomass cookstoves and provide indications for a better testing approach. A laboratory based comparative study of different available protocolswas done,and the impact of various parameters on the performance of stoves was also studied.

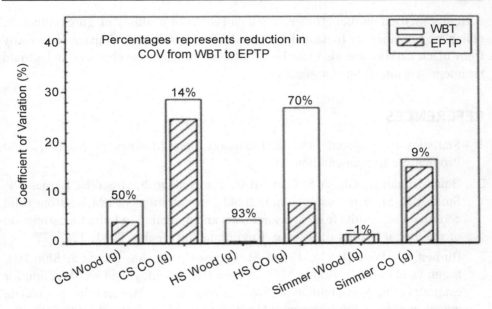

Figure 11.2: Coefficients of Variation for WBT and EPTPMethods [37, 38]

The thermal performance and emission performance were found to be different among each protocol. The differences are as a result of variation in combustion conditions followed in the protocols and the parameters used in calculations. However, the BIS protocol gives a better simulation of Indian cooking cycles. For instance, in terms of the energy in the charcoal left at the end of testing, it is treated as useful energy under WBT protocol and not counted in calculations for thermal efficiency by not including it in the energy delivered by the fuel consumed. While under BIS, this is counted as the energy delivered by the fuel consumed. As compared to WBT, where water is heated in vessels without putting on the lids, the tests under BIS protocol are conducted with lidded vessels that represent the actual cooking practice in India. The lid on the vessel avoids the uncertainties resulting due to evaporation of water. Hence BIS leads to lower error percentage (uncertainty) as compared to the WBT protocol. Similarly, BIS uses a different size of vessels, and the amount of water taken there for different sizes of stoves, as well as the amount of fuel, also varies with the size/capacity of the stove. At the same time, in WBT, it is the same size of vessel and amount of water used for different sizes/capacities of stoves.

The present work tried to cover substantial ground in the area of cookstoves testing methodologies and performance evaluation of any cookstove with the help of protocols, yet, obviously, further studies need to be taken up in the future, particularly in the Climate implications areas. Different cookstoves protocols currently focused on fuel efficiency and emissions of health-damaging pollutants, which are priorities for India and the world. However, Climate impacts, especially from black carbon, are also can be of interest to many stakeholders and should be integrated into future protocols.

REFERENCES

1. Statistics E. National Statistical Organization. Ministry of Statistics and Programme Implementation. 2013.
2. Balakrishnan K, Ghosh S, Ganguli B, Sambandam S, Bruce N, Barnes DF, Smith KR. State and national household concentrations of PM 2.5 from solid cookfuel use: results from measurements and modeling in India for estimation of the global burden of disease. Environmental Health. 2013; 12(1):77.
3. Burnett RT, Pope III CA, Ezzati M, Olives C, Lim SS, Mehta S, Shin HH, Singh G, Hubbell B, Brauer M, Anderson HR. An integrated risk function for estimating the global burden of disease attributable to ambient fine particulate matter exposure. Environmental Health Perspectives. 2014; 122(4): 397-403.
4. Kshirsagar MP, Kalamkar VR. A comprehensive review on biomass cookstoves and a systematic approach for modern cookstove design. Renewable and Sustainable Energy Reviews. 2014; 30: 580-603.
5. Ruiz-Mercado I, Masera O. Patterns of stove use in the context of fuel–device stacking: rationale and implications. Eco Health. 2015; 12(1): 42-56.
6. IEA AE. A focus on the energy prospects in sub-Saharan Africa, World Energy Outlook Special Report. International Energy Agency Publication. 2014;1:e237.
7. Jetter J, Zhao Y, Smith KR, Khan B, Yelverton T, DeCarlo P, Hays MD. Pollutant emissions and energy efficiency under controlled conditions for household biomass cookstoves and implications for metrics useful in setting international test standards. Environmental Science and Technology. 2012; 46 (19): 10827-34.
8. Simon GL, Bailis R, Baumgartner J, Hyman J, Laurent A. Current debates and future research needs in the clean cookstove sector. Energy for Sustainable Development. 2014; 20: 49-57.
9. Johnson M, Edwards R, Masera O. Improved stove programs need robust methods to estimate carbon offsets. Climatic Change. 2010;102(3-4):641-9.
10. Kishore VV, Ramana PV. Improved cookstoves in rural India: how improved are they?: A critique of the perceived benefits from the National Programme on Improved Chulhas (NPIC). Energy. 2002; 27(1): 47-63.

11. Lombardi F, Riva F, Bonamini G, Barbieri J, Colombo E. Laboratory protocols for testing of Improved Cooking Stoves (ICSs): A review of state-of-the-art and further developments. Biomass and Bioenergy. 2017; 98: 321-35.
12. Raman P, Ram NK, Murali J. Improved test method for evaluation of biomass cook-stoves. Energy. 2014; 71: 479-95.
13. Baldwin SF. Biomass stoves: engineering design, development, and dissemination. Arlington, VA: Volunteers in Technical Assistance; 1987.
14. Yuntenwi EA, MacCarty N, Still D, Ertel J. Laboratory study of the effects of moisture content on heat transfer and combustion efficiency of three biomass cook stoves. Energy for Sustainable Development. 2008; 12(2): 66-77.
15. Bhatt MS. The efficiencies of firewood devices (open-fire stoves, chulahs and heaters). Proceedings of the Indian Academy of Sciences Section C: Engineering Sciences. 1982; 5(4): 327-42.
16. MacCarty N, Still D, Ogle D, Drouin T. Assessing cook stove performance: field and lab studies of three rocket stoves comparing the open fire and traditional stoves in Tamil Nadu, India on measures of time to cook, fuel use, total emissions, and indoor air pollution. Aprovecho Research Center. 2008:1-8.
17. Jetter JJ, Kariher P. Solid-fuel household cook stoves: characterization of performance and emissions. Biomass and Bioenergy. 2009; 33(2): 294-305.
18. Bussmann PJ, Visser P, Prasad KK. Open fires: experiments and theory. Proceedings of the Indian Academy of Sciences Section C: Engineering Sciences. 1983; 6(1): 1-34.
19. Still D, MacCarty N, Ogle D, Bond T, Bryden M. Test results of cook stove performance. AprovechoResarch Center, Shell Foundation, United States Environmental Protection Agency. 2011: 126.
20. Bussmann PJ. Woodstoves: theory and applications in developing countries. 1988.
21. Dirks A M. Test on the experimental stove. Clean combustion of wood- Part-2. http:// www.nzdl.org/gsdlmod?e=d-00000-00-off-0fnl2.2—00-0-0-10-0-0-0direct-10-4-0-1l—11-ky-50-20-preferences-10-0-1-00-0-4-0-0-11-10-0utfZz-8-00&cl=CL2.10.1&d=HASH01 d6bfacc02501 ea394888b2.12&gt=1[accessed 27 August, 2020].
22. Bhattacharya SC, Albina DO, Khaing AM. Effects of selected parameters on performance and emission of biomass-fired cookstoves. Biomass and Bioenergy. 2002; 23(5): 387-95.

23. De Lepeleire G, Christiaens M. Heat transfer and cooking woodstove modelling. Proceedings of the Indian Academy of Sciences Section C: Engineering Sciences. 1983; 6(1): 35-46.
24. Mukunda HS, Dasappa S, Paul PJ, Rajan NK, Yagnaraman M, Kumar DR, Deogaonkar M. Gasifier stoves—science, technology and field outreach. Current Science (00113891). 2010; 98(5).
25. Joseph S, Shanahan Y. Designing a test procedure for domestic woodburning stoves. London: Intermediate Technology Development Group; 1980.
26. Bussmann P, Visser P, Delsing J, Claus J, Sulilatu W. Some studies on open fires, shielded fires and heavy stoves. Prasad KK, editor. Eindhoven University of Technology; 1981.
27. Adria O, Bethge JP. Chinese National Improved Stove Program (CNISP).
28. Arora P, Das P, Jain S, Kishore VV. A laboratory based comparative study of Indian biomass cookstove testing protocol and Water Boiling Test. Energy for Sustainable Development. 2014; 21: 81-8.
29. GERES Cambodia Adapted Water Boiling Test (AWBT). http:// cleancookstoves .org/ Binary-data /Document/file/000/000/75-1.pdf [accessed 27 August, 2020].
30. Lombardi F, Riva F, Bonamini G, Barbieri J, Colombo E. Laboratory protocols for testing of improved cooking stoves (ICSs): A review of state-of-the-art and further developments. Biomass and Bioenergy. 2017; 98: 321-35.
31. Berrueta VM, Edwards RD, Masera OR. Energy performance of wood-burning cookstoves in Michoacan, Mexico. Renewable Energy. 2008; 33(5): 859-70.
32. Mac Carty N, Ogle D, Still D, Bond T, Roden C. A laboratory comparison of the global warming impact of five major types of biomass cooking stoves. Energy for Sustainable Development. 2008; 12(2): 56-65.
33. Roden CA, Bond TC, Conway S, Pinel AB, MacCarty N, Still D. Laboratory and field investigations of particulate and carbon monoxide emissions from traditional and improved cookstoves. Atmospheric Environment. 2009; 43(6): 1170-81.
34. Adkins E, Tyler E, Wang J, Siriri D, Modi V. Field testing and survey evaluation of household biomass cookstoves in rural sub-Saharan Africa. Energy for Sustainable Development. 2010; 14(3): 172-85.
35. MacCarty N, Still D, Ogle D. Fuel use and emissions performance of fifty cooking stoves in the laboratory and related benchmarks of performance. Energy for Sustainable Development. 2010; 14(3): 161-71.

36. Zhang Y, Pernberton-Piqott C, Zhang Z, Ding H, Zhou Y, Dong R. Key differences of performance test protocols for household biomass cookstoves. In Twenty-Second Domestic Use of Energy 2014 (pp. 1-11). IEEE.

37. L'Orange C, Volckens J, DeFoort M. Influence of stove type and cooking pot temperature on particulate matter emissions from biomass cook stoves. Energy for Sustainable Development. 2012;16(4):448-55.

38. L'orange C, DeFoort M, Willson B. Influence of testing parameters on biomass stove performance and development of an improved testing protocol. Energy for Sustainable Development. 2012; 16(1): 3-12.

39. Verma AR, Prasad R, Vijay VK, Tiwari R. Modifications in improved cookstove for efficient design. In Proceedings of the First International Conference on Recent Advances in Bioenergy Research 2016 (pp. 245-253). Springer, New Delhi.

40. Trivedi A, Verma AR, Kaur S, Jha B, Vijay V, Chandra R, Vijay VK, Subbarao PM, Tiwari R, Hariprasad P, Prasad R. Sustainable bio-energy production models for eradicating open field burning of paddy straw in Punjab, India. Energy. 2017; 127: 310-7.

41. Mal R, Prasad R, Vijay VK, Ranjan A, Tiwari R. Thermoelectric power generator integrated cookstove: a sustainable approach of waste heat to energy conversion. International Journal of Research in Engineering and Technology (IJRET). 2014; 3: 35-40.

CHAPTER - 12

DENSIFICATION TECHNOLOGIES FOR AGRO WASTE MANAGEMENT

Abolee Jagtap* and Surendra R. Kalbande

*Department of Unconventional Energy Sources and Electrical Engineering
Dr. Punjabrao Deshmukh Krishi Vidyapeeth, Akola-444104, Maharashtra, India*
Corresponding Author

Due to an increase in worldwide population, the energy demand in developing countries is more and the rate of energy production is significantly reduced. Developing nations produce a huge amount of agro residues and some peasants are burning this surplus residue in an open field, which creates excessive pollution, a hazy and smoky environment. The most commonly available agro-waste material includes rice husk, soybean straw, groundnut shells, cotton stalks jute, sugarcane bagasse, and coffee husk, etc. This surplus biomass is facing some inherent problems in handling, storage, and energy generation due to its low bulk density, low calorific value, high moisture content, and hydrophilic nature. Therefore, the appropriate conversion of agro-waste into energy-rich products is became an important hotspot in the coming years. There are various biomass conversion technologies are available but most people carried out the direct burning of the material which produces a very less amount of energy. Therefore, the densification of biomass is considered as one of the promising methodology for the production of briquettes. Biomass briquetting technology is one of the best alternative fuel sources of energy as compared to the direct burning of the crop residues. Densification means compaction of material or to increase the density of loose biomass.

1. INTRODUCTION

Due to an increase in the worldwide population, the global energy demand is also raising which causes depletion of fossil fuels, and simultaneously emission of greenhouse gases is increased in the atmosphere. In the world primary energy is consumed up to 79%, among these 57.7% contribute to the transportation sector which is going to diminish rapidly. Among the different available renewable energies to achieve energy demand, biomass is measured as one of the greatest resources for obtaining green energy. Biomass is a biodegradable organic material, which is produced through the chemical storage of solar energy as a result of photosynthesis. Although due to the availability of a large amount of fossil fuels, most probably biomass is considered as a secondary resource of energy with a very less contribution to energy generation. We all know very well that India is an agricultural prominent country; therefore, a large amount of agricultural production is recorded annually. This results in an increase in the availability of surplus residue mainly known as biomass can be consumed appropriately for the production of other value-added green products [1]. Developing nations produce a huge amount of agro residues while; some peasants are burning this surplus residue in an open field, which creates excessive pollution, a hazy and smoky environment. The most commonly available agro-waste material includes rice husk, soybean straw, groundnut shells, cotton stalks jute, sugarcane bagasse, and coffee husk, etc. This surplus biomass is facing some inherent problems like handling, storage, transportation, and energy generation due to its low bulk density, low calorific value, high moisture content, and hydrophilic nature, etc. Therefore, the appropriate conversion of agro-waste into energy-rich products is becoming an important hotspot in the coming years. There are various biomass conversion technologies are available but most people carried out the direct burning of the material which produces a very less amount of energy. Therefore, the densification of biomass is considered as one of the favourable technology for the production of briquettes.

Biomass densification means compression of loose material or to increase the density of loose biomass. Direct use of biomass for fuel purposes in large scale applications is very difficult because it has low bulk density, high moisture content, and more volume. Biomass briquettes can be made with high density of 1.2 to 1.4 g/cm^3 from loose biomass with a bulk density of 0.1 to 0.2 g/cm^3. Biomass densification technology converts organic waste into a value-added fuel product these are also known as pelletizing, briquetting, or agglomeration which recovers transportation and storage materials. Densification technology has been useful for many nations in developing countries. For biomass densification William Smith was first researcher to be received a United States patent (1880). He was used a steam hammer (at 66°C), and compacted waste from sawmills [2].

Briquettes are prepared from combustible material which is obtained from agricultural raw material, forest waste material, or coal dust from industry, Table 12.1 shows the different agro-waste material used for the briquette production. The biomass densification process can be divided into different categories such as balling, pelletization, extrusion, and briquette which are carried out using bailer, screw press, piston, and roller. Pelletization technology and briquetting technology are the popular process used for compaction of biomass. In the screw press technology, biomass is extruded continuously through a heated die. In the screw press process, product quality and production rates are higher than the piston press process. However, comparing wear parts in a piston press and screw press it was observed that the screw press wear parts require more repairing and operational maintenance. The centre hole appeared on the briquette produced by screw press technology would achieve uniform combustion and the material would be carbonized rapidly due to good heat transfer [4].

Table 12.1: Raw Materials Used for Briquette Production [3]

Sr. No.	Biomass Resources	Raw Materials
1	Agricultural wastes from field	Cassava, coconut frond, corn stalks, straw, millet, oat straw, frond palm oil, rice straw, rye straw, sorghum straw, soybean straw, sugar reed leaves, wheat straw, cotton stalks
2	Industrial Residues	Cocoa beans, coconut shells, coffee husks waste, cottonseed hulls, groundnut shells, maize, oil palm stalks, olive press waste, sugarcane bagasse
3	Forest Waste	Leaves, branches, and trunks.
4	Wood industry wastes	Sawdust

2. MECHANICS OF BONDING OF PARTICLES

The densified biomass product quality depends on the different parameters like die diameter, die temperature, pressure, usage of binders, and preheating of the biomass mix. According to the study, biomass compaction during the densification process can be attributed to elastic and plastic deformation of the particles at higher pressure. Three stages occur during the densification of biomass. During the first stage densification process, biomass particles remain in packed mass, and also physicochemical properties of particles remain as it is while some energy losses occur due to the particle to the wall and inter-particle friction. Followed by, second stage process in which due to inter particles contact undergoes elastic and plastic deformation which might be caused due to Vander Waal's electrostatic forces. Finally, in the case of the third stage process volume was significantly dropped due to the higher pressure while the density of pellet product reaching towards the true density [5]. Figure 12.1 shows the mechanism of loose material under compression [2].

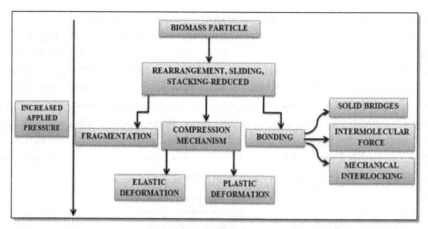

Figure12.1: Mechanism of Loose Material under Compression [6,7]

At the end of the third phase, the broken and deformed particles are no longer changed their position because of a reduction in cavities. There are different variables such as a combination of temperature, process equipment, and pressure significantly affect the densification process and govern the performance of the system. However, if theses parameters not optimized carefully, then it influences a negative effect on the intra-particle cavities and also on the conversion process.

3. RAW MATERIALS FOR BRIQUETTING AND PRE-TREATMENTS

3.1 Collection of Materials

Generally, any material will burn, but it is not in proper shape, size or form to be readily used as fuel is considered as a good material for briquetting. The most common agro residues used for the densification process are soybean straw, cotton straw, sugarcane bagasse, cow dung, rice husk, leaves branches, sawdust etc. Figure 12.2 shows the transportation of biomass material from the field.

3.2 Pre-treatment of Raw Materials

The pre-treatment process of biomass samples is categorized as sorting, drying, and finally size reduction.

Sorting: The raw material is carefully separated manually to remove impurities such as pieces of wood, bone, metal, and any other unwanted materials.

Drying: The virgin biomass contains more moisture content therefore it is necessary to minimize the moisture content from biomass, which may result in a reduction of biomass weight. This method reduces processing costs, as well as storage and transport cost. The dried end product is used as a plant nutrient or as a fuel. Drying can be occurring in open sun drying or solar dryers with heater or with hot air. The drying of biomass in the solar dryer is shown in Fig. 12.3.

Figure12.2: Transportation of Biomass Material from the Field [8]

Figure 12.3: Drying of Biomass in the Solar Dryer [9]

Size Reduction: The main motive of size reduction of biomass is to convert the material which is optimizing for transportation, handling, and storage. In addition, the small particle process that is helpful for pelletization, it increases the bulk density of the material, and develops the conversion process [8]. Size reduction processes can be categorized as 1) Single breaking process such as shredding, cutting, shearing as roller grinders, forage choppers, shredders, crushers and rotary veneer choppers; and 2) Multiple breaking milling process (Fig. 12.4) such as knife mills, ball/rod mills, disk (attrition) mills, hammer mills, and ultrafine mills [9].

Figure 12.4: Some Size Reduction Machinery [9]

4. DENSIFICATION TECHNOLOGIES

4.1 Piston Press Type Machine

Piston press type machine is the most commonly used densification technology for the compaction of biomass. In the piston press machine, the material is fed through hopper into the cylinder which is then compacted by the reciprocating motion of the piston towards the tapering die as shown in Fig. 12.5. In the market, 0.5-1.5 t/h capacities of machines are available. They are produce 50-90 mm diameter briquettes. Piston presses can be operating mechanically or hydraulically. Mechanical piston press generally produces durable and compact briquettes, while hydraulic press which works at lower pressure, produced briquettes that are less dense, soft and friable. The mechanical press produces force approximately 2000 kg/cm^3 without the addition of binder. Energy loss in a mechanical machine is less and power consumption is optimal. The power requirement of briquetting machines varies from 25 to 66 kW. The operating life of the mechanical press is longer than the hydraulic press. Generally, a mechanical press gives a better return rate on investment than a hydraulic press [2].

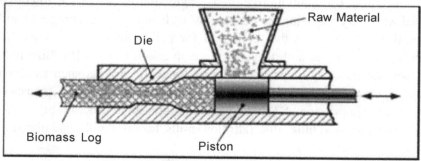

Figure 12.5: Piston Press Machine [10]

4.2 Screw Press Type Machine

In the screw type of machine, the material is fed via a hopper, which is continuously moving forwards the cylindrical die due to external drive applied on the screw as shown in Fig. 12.6.

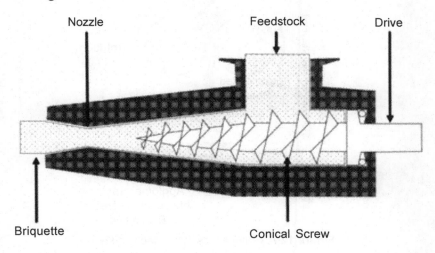

Figure12.6: Screw Press Type Machine [11]

During screw extrusion, biomass passes from the feed port, with a rotating screw, through the barrel and against the die, which causes a similar pressure gradient and wall friction due to the internal shearing of the biomass. The temperature of biomass increases due to friction at the barrel, internal friction due to material, and high rotational speed (~600 rpm) of the screw. If the generated heat inside the system is not enough to reach a pseudo plastic state for smooth extrusion, then external heat could be provided to achieve desired process temperature between 250°C to 300°C where lignin starts flowing. Here, in this technology pressure builds up very smoothly, and therefore screw press type technology is more promising for good quality briquettes production as compared to piston press. Although, it required more maintenance and operational cost due to higher wear and tear inside the machine [12].

4.3 Roller Press Type

In the roller press machine densification system, pressure is applied horizontally in opposite directions on parallel axes and it works on the principle of pressure and agglomeration as shown in Fig. 12.7.

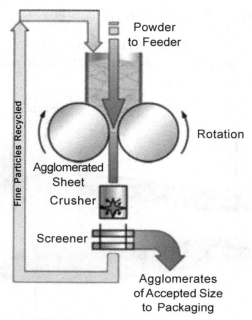

Figure 12.7: Roller Press Type Machine [2]

The roller press machine is consisting of hoppers, feeders, two rollers with the same diameter, crusher, and screener. When biomass enforced through the hole between the two rollers, is pushed into small pockets, the densified product form. Due to rotary rotations of rollers is rotated in opposite directions and on one side biomass is moved and the densified biomass is discharged out from the reverse side. Design parameters that play a most important role for determination of densified product quality such as roller diameter, the gap width, the roller force, and the shape of the die [13]. The bulk density ranges from 450 to 550 kg/m^3.

4.4 Manual Press and Low Press Machine

There are various types manual press machines are available for the densification of biomass feed stocks. Manual clay brick making press is a widely used manual machine. This machine is capable for both raw feedstock and charcoal. The main benefits of low-pressure machine are a minimum investment, low operating costs, and don't require a skilled person to control the technology. Low-pressure machine is used for briquetting green plant waste such as bagasse (sugar-cane residue). The developed briquette has a higher density than the raw material but requires drying for moisture removal before its further thermal application. The dried briquettes are very less mechanically durable but crumble easily.

4.5 Flat Die Type Machine

A pellet mill contains holed flat die with one or two rollers. When die and roller rotated, the feedstock is filled through the holes to form densified pellets. These machines operate by small diameter die (10-30 mm) pellet, which has countable holes as shown in Fig.12.8. The pellet machines are available with production capacities ranging from 200 kg/h to 8 ton/h. The pellet mill power consumption ranges from 15-40 kWh/ton. Pellets are less strong than briquettes [14].

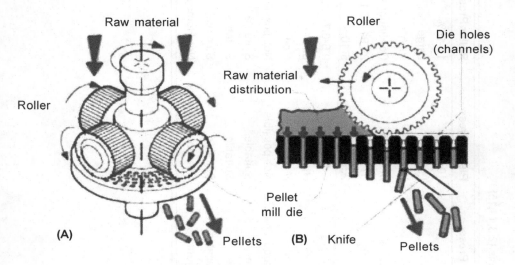

Figure 12.8: Flat Die Type Machine [1]

5. COMPARISON OF DIFFERENT DENSIFICATION TECHNOLOGIES

The different densification processes have been compared in terms of feedstock properties, machine capacity, maintenance, specific energy, and its end product applications as shown in Table 12.2. The screw press densification process is largely suitable for co-firing. The pellet press, roller, and piston press densification processes are more appropriate for the thermo chemical conversion processes of biomass. Piston press densification process shall be preferred for the large particle size of raw materials with higher moisture content [2]. The energy requirement for the densification process mainly depends on several factors it may be for compression, pushing, shearing, and mixing of various raw materials. In the densification process, energy consumption is more for compression and pushing material while the remaining energy is utilized to reduce the frictional losses

Table 12.2: Comparison of Different Densification Systems [2,3,14,15]

Sr. No.	Parameters	Screw Press	Piston Press	Roller Press	Pellet Press
1	Optimum moisture of biomass	8-9%	10-15%	10-15%	10-15%
2	Particle Size	Smaller	Larger	Larger	Smaller
3	Wear of contact part	High	Low	High	High
4	Machine Output	Continuous	In-stroke	Continuous	Continuous
5	Specific energy consumption (kWh/ton)	36.8-150	37.4-77	29.91-83.1	16.5-74.5
6	Through puts (ton/hr)	0.5	2.5	5-10	5
7	Density of briquette (g/cm^3)	1-1.4	1-1.2	0.6-0.7	0.7-0.8
8	Maintenance	Low	High	Low	Low
9	Combustion of briquettes	Very good	Moderate	Moderate	Very good
10	Carbonization of charcoal	Make good Charcoal	Not possible	Not possible	Not possible
11	Suitability in gasifier	Suitable	Suitable	Suitable	Suitable
12	Suitability for co-firing	Suitable	Suitable	Suitable	Suitable
13	Suitability for biochemical conversion	Not-Suitable	Suitable	Suitable	Suitable
14	Homogeneity of densified biomass	Homogenous	Not-Homogenous	Not-Homogenous	Homogenous

meanwhile the compression process. Screw press technology consumes higher energy due to more wear and tear occurred in the system, on the other hand, pellet press consumes less amount of energy which may be due to the suitability of high moisture feedstock. Therefore, pellet press biomass briquettes possess very little durability as compared to other densification technology [15].

6. CHARACTERISTICS OF BIOMASS BRIQUETTES

6.1 Physical Properties

The physical properties of briquettes include moisture content, length, width, diameter, bulk density, shatter resistance of briquettes, tumbling resistance of briquettes, and water penetration resistance are given below:

6.1.1 Moisture Content

The briquettes moisture content can be determined by using oven-dry method. The initial sample briquette with known weight is kept in an oven at 105°C for one hour. Then the oven-dry sample weighted using electric weighing balance. The moisture content of the sample calculated by using the following formula:

$$\text{M.C.} = \frac{W_2 - W_3}{W_2 - W_1} \times 100 \qquad \ldots (1)$$

Where, W_1 = weight of crucible, g
                W_2 = weight of crucible + sample, g
                W_3 = weight of crucible + sample, after drying, g

6.1.2 Bulk Density

For storage and transportation purpose biomass bulk density is an important parameter. The high density of briquettes is preferred as fuel because of more energy content as compared to raw biomass per unit volume and less-burning property [16]. The bulk density of briquettes is evaluated by using standard method. A cylindrically shaped container is used for determination with volume 1000 ml. The unfilled container is weighed to determine its mass and after weighed it is filled with the briquette sample and weighed. The bulk density is determined by dividing the mass of the material by the volume of the container [17]. The bulk density is calculated by using the formula:

$$\text{Bulk Density (kg/m}^3\text{)} = \frac{\text{Mass of volume sample (kg)}}{\text{The volume of measuring cylinder (m}^3\text{)}} \quad \dots (2)$$

6.1.3 Shatter Resistance Test

Shatter resistance test is used for determination of hardness of the briquettes. The known weight and length of briquette is fallen from the one-meter height on a floor for ten times. The weight of briquette and its size is recorded. The loss of material is calculated. The shatter resistances of the briquette are calculated by using the following formula [18].

$$\text{Per cent weight loss} = \frac{W_1 - W_2}{W_1} \times 100 \quad \dots (3)$$

$$\text{\% shatter resistance} = 100 - \text{\% weight loss} \quad \dots (4)$$

Where,

W_1 = weight of briquette before shattering, g
W_2 = weight of briquette after shattering, g

6.1.4 Tumbling Resistance Test

This test is used for determining the durability of products. The durability index is one of the most important parameters which defined as the ability of briquettes materials to remain undamaged when handling throughout storage and transportation. The briquette durability is the physical strength and resistance to being shattered up. This test is determined by keeping a known quantity of pellets in an airtight metallic box. The box is twisted round to rub the pellets one another. The powder is generated and weighed by using weighing balance. Less generated dust means better durability of the pellets [19]. Different types of equipment (tumbling can, Ligno tester, Holmen tester, and Dural tester) are used to durability test [15].

$$\text{Per cent weight loss} = \frac{W_1 - W_2}{W_1} \times 100 \quad \dots (5)$$

$$\text{Tumbling resistance} = 100 - \text{\% weight loss} \quad \dots (6)$$

Where,

W_1 = weight of briquette before tumbling, gm
W_2 = weight of briquette after tumbling, gm

6.1.5 Water Penetration Resistance

When biomass briquette is immersing in water at room temperature for 30 seconds, it records the percentage of water gained by a briquette. The water absorption percentage is calculated by using equation no 7 and water penetration resistance is calculated by equation 8 as below [20].

$$\text{Per cent weight loss} = \frac{W_1 - W_2}{W_1} \times 100 \quad \ldots (7)$$

% resistance to water penetration = 100 - water gain ... (8)\

Where,

W_1 = Initial weight of briquette

W_2 = final weight of briquette.

6.1.6 Degree of Densification

It is defined as a percent density of biomass increases due to the densification process. The degree of densification measured the capability of a material to get bounded. It is recorded by using the equation below [20].

$$\text{Degree of densification} = \frac{\text{Density of briquette - Density of raw materials}}{\text{Density of raw material}} \quad \ldots (9)$$

6.1.7 Energy Density Ratio

The energy density ratio means ratio of energy content as per unit volume of raw material and the energy content as per unit volume of briquette fuel. The energy density ratio of briquette fuel is calculated by using the equation below [21].

$$\text{Energy density ratio} = \frac{\text{Energy content of briqetted fuel (kcal/m}^3\text{)}}{\text{Energy content of raw materials (kcal/m}^3\text{)}} \quad \ldots (10)$$

6.2 Thermal Properties of Briquettes

The thermal properties are the most important parameter of briquette. They include volatile matter, calorific value, ash content, and fixed carbon [21]. Proximate analysis of the briquette is evaluated the percentage of volatile matter, moisture content, ash content, and fixed carbon.

6.2.1 Volatile Matter

The volatile matter is determined using the standard method. 2 gm of powder briquette sample is taken and oven-dried at 105°C till its mass is constant. After the sample is heated at 550°C for 10 min and weighted the cool sample. The volatile matter is determined using the following formula [21]:

$$\text{Volatile matter} = \frac{A-B}{A} \times 100 \qquad \ldots (11)$$

Where,

A = Weight of the oven-dried sample

B = eight of the sample after 10 min in the furnace at 550°C

6.2.2 Ash Content

The sample is kept in crucible and heated in a muffle furnace without lid at 700 ± 50°C for a one-half hour. After that crucible is taken out, cooled down in air. Then kept in desiccators and weighed. The process is repeated, till constant weight is obtained. The following formula is used:

$$\text{Ash content (\%)} = \frac{W_5 - W_1}{W_2 - W_1} \times 100 \qquad \ldots (12)$$

Where,

W_1 = weight of empty crucible, g

W_2 = weight of crucible + sample taken from the stage (II), g

W_5 = weight of crucible + ash left in a crucible, g

6.2.3 Fixed Carbon

The fixed carbon content is determined by using the mass balance equation for the biomass sample. The following equation is used:

$$FC\ (\%) = 100 - \%\ \text{of}\ (MC + VM + AC) \qquad \ldots (13)$$

Where,

FC = Fixed Carbon

MC = Moisture Content

VM = Volatile Matter,

AC = Ash Content

6.2.4 Calorific Value

The calorific value of briquette is determined by using a bomb calorimeter method. The calorific value of briquette is determined by using the following formula [21].

$$\text{Calorific Value} = \frac{(W+w) \times (T_1 - T_2)}{X} \quad \ldots (14)$$

Where,

W = weight of water in calorimeter (kg)
w = water equivalent of apparatus
T_1 = initial temperature of water (°C)
T_2 = final temperature of water (°C)
X = weight of fuel sample taken (kg)

6.3 Energy Evaluation Analysis

The evaluation of the energy content of briquettes is calculated on the basis of thermal fuel efficiency, burning rate, and ignition time.

6.3.1 Thermal Fuel Efficiency (TFE) Test

In steel vessel 100 litres of water placed for the determination of the thermal efficiency of fuel during the experiments. The vessel is accurately closed to reduce evaporation losses and engaged on a biomass stove. The vessel temperature is recorded by using thermometer. The 5 kg briquettes samples is recorded and separated into four parts for experiments. The ambient water temperature in the jar is measured and after that briquettes are ignited.

The boiling water temperature is recorded. The water is boiled up to evaporation of water vapour and then briquettes are completely burned. The lid is removed from container and for 20 min water is evaporated continuously. The container is removed from the stove and kept for cooling 2 h. The remaining volume of water is recorded. Using a stopwatch, the time between T_0 and T_b is measured. The thermal efficiency of densified briquettes is calculated using the following formula [22]:

$$\text{Thermal Fuel Efficiency} = \frac{M_w \times C_p \times (T_b - T_o) + M_c L}{M_f E_f} \quad \ldots (15)$$

Where,

M_w = mass of water (kg)
C_p = specific heat of water (kJ/kg K)
T_b = boiling temperature of water (K)
T_0 = initial temperature of water (K)
M_c = mass of water evaporated (kg)
L = latent heat of evaporation (kcal/kg)
M_f = mass of fuel burnt (kg)
E_f = calorific value of fuel (kJ/kg)

6.3.2 Burning Rate

The burning rate means the rate at which a particular mass of fuel is burnt in air. Some amount of the briquette is kept on the wire gauze and the gas burner burned. The weight of the gauze is continuously recorded up to the sample is completely shattered and constant weight take place. The weight loss at a particular time is calculated [23]:

$$B_s = \frac{Q_1 - Q_2}{T} \quad \ldots (16)$$

Where

B_s = Burning rate (g/min)
Q_1 = Initial weight of fuel prior to burning (g)
Q_2 = Final weight of fuel after burning (g)
T = Total burning time (min)

6.3.3 Ignition Time

The ignition time meanstime taken to ignite the known weight of biomass briquettes Exactly 100 g of briquette is placed on a wire mesh grid (of known mass resting) in between two fire-retardant bricks to allow free flow of air around it. A Bunsen burner is placed directly underneath this platform and adjusted to a blue flame. The burner is lighted until the briquette is ignited. The ignition time is computed using the formula below [24]:

$$\text{Ignition time} = t_1 - t_o \quad \ldots (17)$$

Where

t_1 = time the briquette ignited (min),

t_0 = time the burner was lighted (min).

7. ECONOMICS OF BRIQUETTING TECHNOLOGY

The briquettes are used for energy generation helping farmers to earn money from the waste. Briquetting of residues takes place with the application of pressure, heat and on the loose materials to produce the briquettes. Economics of biomass briquetting machine is depends on site location and local condition of regions. The basic economic aspects of briquettes machine considered in local condition. The technology requires high energy power or low energy power. Commercially piston press machine is mostly used as compared to screw press machine. In the screw press machine wear of contact part is high and low in piston press. Due to high wear of contact parts in screw press, the power consumption is more near about 60 kWh/ton and 50kWh/ton in piston press. Also the installation of capital cost is less for piston press machine compared to screw press [18]. Kanagaraj et al., 2017 reported that total fixed cost of briquette production was Rs. 341 per ton, fixed cost and interest on working capital (Rs.128 per ton), followed by depreciation on machinery (Rs. 65.45 per ton) and rent paid for land (Rs. 48.00 per ton) as shown in Table 12.3.

Table 12.3: Fixed Cost in Per Ton of Briquette Production [25]

Sr. No.	Particulars	Rs per ton
1	Salary to staff	39.00
2	Rent paid for land	48.00
3	Electricity for installation	30.00
4	Depreciation of building (3.34%)	5.93
5	Depreciation of machinery (11.3%)	65.45
6	Tax (2% on total investment)	25.00
7	Interest on capital	128.00
8	**Total fixed cost**	**341.00**

The variable cost of briquette production per ton was 3100. The cost of raw material alone accounted for Rs. 2300 per ton. The cost of man power and electricity were 240 Rs/ton and 230 Rs/ton (7.45 %) respectively. The storage cost of briquette was Rs. 30 per ton. In total cost of production, the fixed cost and variable cost were recorded Rs.341 per ton and Rs. 3100 per tonne respectively as shown in Table 12.4.

Table 12.4: Variable Cost in Rs/ton of Briquette Production [25]

Sr. No.	Particulars	Rs per ton
1	Cost of raw material	2300.00
2	Transportation cost	180.00
3	Cost of electricity	230.00
4	Man power	240.00
5	Repair and maintenance	70.00
6	Stationary and supplies	50.00
7	Cost of Storage	30.00
8	**Total variable cost**	**3100.00**
9	**Cost of production per ton**(Fixed cost + Variable cost)	**3441.00**

The gross return generated per tonne of briquette is arrived at Rs 4700 and generated a net income of Rs. 1229 per ton of briquette after deducting the cost of production involved per tonne of briquette as shown in Table 12.5.

Table 12.5: Returns from Briquette Production [25]

Sr. No.	Particulars	Rs per ton
1	Gross return	4700.00
2	Total cost	3441.00
	Net return per tonne	1229.00

8. BINDERS USED IN BIOMASS DENSIFICATION TECHNOLOGY

Binders are the most important parameter during the densification process. It improves the quality of briquettes, binding properties, and produces more long-lasting products. Binders reduce the wear in manufacture equipment and increase the scrape resistance of the fuel. In the densification process, binders are mostly used but it depends on the final product. The addition of some binders can be an increase in the sulphur content of densified biomass. The most commonly used binder is Lignosulfonates (Wafolin), or sulfonate salts made from the lignin in pulp mill liquors. The following subsection describes the various types of binders used for biomass densification process

8.1 Lignosulfonates

Lignosulfonates have commonly used binders in animal feeds and have been considered the most effective and popular binders [2]. The composition of

Lignosulfonates is sulfonate salts made from lignin of sulphite pulp mill liquors. The general levels of inclusions for effective binding are 1-3%.

8.2 Bentonite

Bentonite means colloidal clay, is commonly used as a binder in feed biomass pelleting and is prepared fromaluminium silicate composed of montmorillonite. During processing ofdensified biomass, gel is form from binder with water to improve the binding properties. Some researchers report that the addition of bentonite with rate of 100 kg/ton of feed mash simultaneously improves the durability of feed material consisting of ground sorghum grain, ground yellow corn and soymeal ingredients.

8.3 Starches

Starches are most commonly used product in the food industry as a binder. The researcher reports that during pelletization pre-cooked starch work as a good binder.

8.4 Protein

Proteins binder are natural binders that are stimulated through the heat produced in the dies. The protein interacts with the other biomass compositions such as lipids and starches, which results in the formation of protein starch and lipid complexes which helps in producing more durable pellets. Alfalfa biomass has high protein content and which can be used as a binder to improve the durability of pellets which is made from lower protein biomass.

9. CONCLUSIONS

Among all technologies discussed in this book chapter, the pellet machine and briquette machine are the most popular and commonly used technologies. The final product of the biomass densification process is applicable for the biochemical, thermo chemical, and co-firing process. The energy consumption in the screw press mill is more while in pellet mill is less. The addition of binders in the case of natural or commercial is also a good alternative to improve the binding properties of pellets. The most important characteristics of biomass briquettes include moisture content, durability index, proximate analysis, bulk density, and calorific value are calculated using the standard method.

REFERENCES

1. Jack H. Flat die and ring die pellet mills comparison, http://www.biofuelmachines. com/flat-die-and-ring-die-pellet-mills-comparison.html [accessed 07 August, 2020].
2. Tumuluru JS, Wright CT, Hess JR, Kenney KL. A review of biomass densification systems to develop uniform feedstock commodities for bioenergy application. Biofuels, Bioproducts and Biorefining. 2011; 5(6): 683-707
3. FAO, Bioenergieetsecuritealimentaire evaluation rapide (BEFS RA), Manuel d'utilisation (Briquettes). Powder Technology. 2014; 127: 162-172
4. Sokhansanj S, Mani S, Bi X, Zaini P, Tabil L. Binderless pelletization of biomass. American Society of Agricultural and Biological Engineers. 2005; 1-13.
5. Mani S, Tabil LG, Sokhansanj S. Compaction characteristics of some biomass grinds. AIC 2002 Meeting, CSAE/SCGR Program, Saskatoon, Saskatchewan, 2002; 14-17.
6. Comoglu T. An overview of compaction equations, Journal of Faculty of Pharmacy, Ankara, 2007; 36(2): 123-133.
7. Denny PJ. Compaction equations: a comparison of the Heckel and Kawakita equations. Powder Technology. 2002; 127(2): 162-72.
8. Mani S, Tabil L, Sokhansanj S. Grinding performance and physical properties of wheat and barley straws, corn stover and switch grass. Biomass and Bioenergy. 2004; 27: 339–352.
9. Miao Z, Grift TE, Ting KC. Size reduction and densification of lignocellulosic biomass feedstock for biopower, bioproducts, and liquid biofuel production. Encycl. Agric. Food Biol. Eng. 2014; 1-4.
10. Chen Z, Yu G, Yuan X, Wang Q, Kan J. Improving the conventional pelletization process to save energy during biomass densification. Bioresources. 2015; 10(4): 6576-6585.
11. Kpalo SY, Zainuddin MF, Manaf LA, Roslan AM. A review of technical and economic aspects of biomass briquetting. Sustainability. 2020; 12(11): 1-30.
12. Victor VM, Jogdand SV, Chandraker AK. Biomass densification technologies to obtain briquettes for energy application-a review. International Journal of Engineering Research and Technology. 2015; 20(3): 1-4.
13. Yehia KA. Estimation of roll press design parameters based on the assessment of a particular nip region. Powder Technology. 2007; 177(3): 148-153.

14. Grover PD, Mishra SK. Biomass briquetting: technology and practices. Bangkok, Thailand: Food and Agriculture Organization of the United Nations. 1996; 46.

15. Kaliyan N, Morey N. Densification characteristics of corn stover and switchgrass, Annual International Meeting, Portland, ASAE, St. Joseph, MI, USA. 2006; 9-12.

16. Kumar R, Chandrashekar N, Pandey KK. Fuel properties and combustion characteristics of Lantana Camara and Eupatorium Spp. Current Science. 2009; 97(6): 930-935.

17. Karaosmanoglu F, Tetik E, Gurboy B, Sanli I. Characterization of the straw stalk of the rapeseed plant as a biomass energy source. Taylor and Francis Group, Energy Sources. 1999; 21: 801-810.

18. Madhava M, Prasad BV, Koushik Y, Rameshbabu KR, Srihari R. Performance evaluation of a hand operated compression type briquetting machine. Journal of Agricultural Engineering. 2012; 49(2): 46-49

19. Tayade S, Pohare J, Mahalle DM. Physical and thermal properties of briquettes by piston press and screw press. International Journal of Agricultural Engineering. 2010; 3(2): 223-227

20. Birwatkar VR, Khandetod YP, Mohod AG, Dhande KG, Source OE, Dapoli T, Machinery F, Dapoli T. Physical and thermal properties of biomass briquetted fuel. Ind. J. Sci. Res. and Tech. 2014; 2(4): 55-62

21. Sengar SH, Mohod AG, Khandetod YP, Patil SS, Chendake AD. Performance of briquetting machine for briquette fuel. International Journal of Energy Engineering. 2012; 2(1): 28-34

22. Panwar NL, Rathore NS. Design and performance evaluation of a 5 kW producer gas stove. Biomass Bioenergy. 2008; 32: 1349-1352.

23. Ndirika VIO. Development and performance, evaluation of a baking oven using charcoal as source of energy. Nigeria J. Renew. Energy. 1998; 12: 83-91.

24. Onuegbu TU, EkpunobiUE, Ogbu IM, Ekeoma MO, Obumselu FO. Comparative studies of ignition time and water boiling test of coal and biomass briquettes blend. IJRRAS. 2011; 7: 153-159.

25. Kanagaraj N, Sekhar C, Tilak M, Palanikumaran B. Cost and returns of briquette production in Tamil Nadu, India. Int. J. Curr. Microbiol. App. Sci. 2017; 7: 1238-1242.

CHAPTER - 13

RECENT ADVANCEMENT IN BIOCHEMICAL CONVERSION OF LIGNOCELLULOSIC BIOMASS TO BIOETHANOL AND BIOGAS

Sweety Kaur[1], Richa Arora[2] and Sachin Kumar[3*]

[1]Nestle India Limited, Rajarhat, Kolkata-700156, West Bengal, India
[2]Department of Microbiology
Punjab Agricultural University, Ludhiana-141004, Punjab, India
[3]Biochemical Conversion Division,
Sardar Swaran Singh National Institute of Bio-Energy,
Kapurthala-144601, Haryana, India
*Corresponding Author

The major parameters responsible for exploration of substitutive resources of energy are escalating energy consumption rate, enhancement in fossil-fuel requirements, growing prices of fossil fuels and accelerating CO_2 and green house gases level in upcoming years. Although availability of solar, wind, hydrothermal and geothermal energy resources is crucial in generating electricity, however, they are not capable enough to sustain and cope up the requirement of transportation fuels. Hence, this emerges the demand of lignocellulosic biomass in prominence to maintain the provision of various forms of solid, liquid and gaseous biofuels. Several research methodological based-studies have evidenced that it can be feasible to produce a diverse range of bioenergy from lignocellulosic biomass residues, however, with non-competitive prices in concerning to petrochemicals and other renewable energy sources. Bioethanol production from lignocellulosic biomass through biochemical fermentative pathways is a best way to replace the requirement of crude oil (fossil-fuels) and decrease the environmental pollution by providing an alternative transparent fuel which also upgrades the combustion efficiency. Production of biogas through anaerobic digestion of lignocellulosic residues through hydrolytic enzymes produced by consortium of hydrolytic microbes such as *Bacteriocides, Clostridia, Bifidobacteria, Enterobacteriaceae* and *Streptococci* also helps in reducing air pollution, emission of greenhouse gases and contributes in eco-friendly supply of bioenergy. This chapter focuses about different microorganisms involved in production of bioethanol and biogas; various lignocellulosic residual feedstocks utilized by microbes for biofuel production, their biosynthesis process with relation to industrial level settings and encountered challenges while processing with their possible solutions to overcome such challenges.

1. INTRODUCTION

With upcoming years, the global population is estimated to attain about 8.5 billion by the year 2030 which can be protruded forth to 9.7 billion by the year 2050 and relatively, the worldwide need and utilization of energy elevates annually by the growth of 1.3 % till the year 2040, however still below the striking growth of 2.3 % observed in 2018 [1]. The requirement of energy and renewable resources is gradually increasing at global level and the issue of rapid depletion of fossil fuels is impending [2]. The wide-ranging subordination up on non-sustainable fossil fuels to acquire the existing demand of energy shall not be renewed for prolonged period dueto depletion of fossil fuel resources and the impact of this subordination is noticeable in the increment of prices of fossil fuels in the decade prior along with harsh atmospheric effects such as climatic variations. Approximately, merely with 23.7% exhaustion of sustainable bioenergy resources for their requirement, it gets important that the planet should shifts to various existing conventional substitutes of renewable energy and biofuels, expeditiously [3]. Such constraints have brought upon new opportunities that primarily emphasizes on the different approaches for management of energy expenditure and substitutive resources of biofuels in order to improve the efficacy and to decline the emission of greenhouse gases appropriately [4].

In consideration of this upcoming need of energy, various forms of biofuels like bioethanol, biogas and biodiesel are been employed as an alternative for existing fossil fuels like diesel fuels and gasoline in several countries, hence depicting that biofuels can be utilized as alternative sustainable energy resource [5]. Each year about 1.3 billion tons of edible residues comprising of one-third of the universal food generated is disposed on global scale irrespective of any additional application and because of prompt economical broadening and constant rise in human population in developing countries in Asia, agricultural and food residual matter is getting produced in a accelerating pattern. Bioenergy acquired from different substrate resources like agricultural crop residues, forest trees, etc can be a major explication to these issues and employed for generation of heat, electricity and biofuels for transportation with reduced intensity of emission of greenhouse gases (GHGs) in comparison to fossil fuel resources. Lignocellulose is a complex chemical structure that comprises of cellulose (40-50%), hemicelluloses (25-35%) and lignin (15-20%) that can be converted into high energy-rich constituents for their application as biofuel and biochemical by-products. A biofuel is defined as an output of complete fermentation of complex sugar structures through aerobic or anaerobic microbes rather than a traditional non-renewable fuel source derived from crude organic matter like petrol and coal due to environmental disturbances. Human population is now inclining towards more novelistic and new techniques for energy production with requisition of different alternatives of non-renewable fossil fuels. The main producers and consumers of biofuels are countries of Europe, Asia and America [6].

There are several different thermo-chemical and biological techniques that have been applied in order to convert the residual biomasses into different forms of liquid and gaseous fuels. Hence, biofuels can be characterized into two main groups: (i) liquid biofuels derived from different substrates are enormous alternative resources of energy to replace the conventionally existing liquid fossil fuels (like diesel and petrol), thereby exhibiting identical properties of gasoline and other petrol-based fuels, comprising of bioethanol, biobutanol, biomethanol and biodiesels; whereas (ii) gaseous biofuels are derived from the breakdown of residual substrates in the absence of atmospheric oxygen, comprising of biomethane, biohydrogen and biohythane [7]. Bioethanol is one among the major cost-effective sustainable fuel, which is vitaland world-wide demanded because of its commercial applications in broad range in industries that has been produced in order to decrease the gradually escalating air pollution through fossil-fuels. Thus, development of bioethanol from economical and natural residual substrates from various food processing industries and agricultural fields is effective to decrease the utilization of existing fossil fuels in global aspects [8]. Biogas is primarily composed of methane and CO_2 that is generated by a biological process, degrading the organic matter in absence of oxygen through anaerobic digestion technique that help dairies in development and confining of the produced biogas, further implementing in generation of electricity for on-farm operations. This chapter investigates about different microorganisms involved in production of bioethanol and biogas; various lignocellulosic residual feedstocks utilized by microbes for biofuel production, their biosynthesis process with rest to industrial level settings and encountered challenges while processing with their possible solutions to overcome such challenges.

2. MICROORGANISMS INVOLVED IN BIOFUEL PRODUCTION

A combination of culture containing two or more ethanologenic-producers or microbes are usually employed in order to ferment the lignocellulosic (cellulose/hemicellulose/lignin) substrates into bioethanol (Table 13.1). Few primitive ethanologenic microorganisms applied in the fermentation process are *Saccharomyces cerevisiae, Esccherichia coli, Zymomonasmobilis, Schizosaccharomycespombe, Pichiastipis, Candida shehatae, Candida brassicae, Fusariumoxysporum, Mucorindicus, etc.* exhibit an important purpose for production of fermentive ethanol. During fermentation of ethanol, two distinctive energy generating pathways are involved i.e., hexose glucose gets consumed by oxidative-metabolic pathway that renders microbial growth and in fermentative-metabolic pathway fermentation of ethanol occurs [9].

Table 13.1: Comparative Analysis of Facultative-Anaerobic, Ethanol-Producing Microorganisms [14]

Parameters	*Saccharomyces cerevisiae*	*Zymomonas mobilis*	*Escherichia coli*
Nature	Eukaryotic microbe	Gram-negative	Gram-negative
Metabolic pathway	Glycolytic Pathway	Entner-Duodoroff (ED) pathway	Glycolytic Pathway
Ethanol yield (Theoretical)	90 to 93%	98%	88%
Ethanol tolerance (v/v)	15%	16%	6%
Range of pH	2 to 6.5	3.5 to 7.5	4 to 8
Genome size (Mb)	12.12	2.14	5.15

It is to be recommended to carry out aerobic and anaerobic fed-batch operations in a combined manner to improve production and to attain high ethanol yield. *Saccharomyces cerevisiae* is generally utilized for fermentation process of ethanol by extracting starch and polysaccharides from plants as a initial biomass, because they utilizes a very unique scale of substrates for glycolytic pathway and its rate is controlled by the presence of the amount of dissolved oxygen [10]. One main issue that afflicts the ethanol fermentation by yeast *S. cerevisiae* is that when complex sugars are fermented through glycolytic pathway, they generates ATP molecules that are forth utilized by yeast cells in their metabolism for the multiplication and growth. Development of such huge amount of expendable cellular biomass is considered as undesirable downstream product at industrial level, depicting the amount of spent carbon and energy sources that could have been incorporated for enhanced production of ethanol [11]. However, this issue can be solved up to some extent through yeast cell immobilization that reduces its growth, as a result accumulation of ATP molecules increases rapidly which impacts adversely the complete glycolysis pathway, therefore, declining the ethanol production. Other yeast-related issues includes detrimental consequence of atmospheric and sub-cellular distress that gathers in the fermentative settings, leading to enhanced ethanol and sugar concentration that ultimately accelerates the osmotic pressure upon cells, elevated temperature of about 38°C, and other conditions. Such parameters can usually be a major origin of stress when working in a combined way, thereby forming the circumstances much terrible for *S. cerevisiae* and this generated stress retards the overall feasibility of yeast cell, hence hindering its capability to generate ethanol [11].

Such productivity related issues can be conquered commercially by incorporating different microbes that can carry out the fermentation process firmly. Among them, a chief competitor is bacterium *Z. mobilis* which is a facultative anaerobe, gram negative in nature, commercially beneficial due to the ability of its members to disrupt glucose molecule by Entner-Duodoroff (ED) pathway, where one ATP molecule is generated, irrespective of glycolysis where two molecules of ATP are produced, but identically producing two pyruvate molecules which is further decarboxylases into acetaldehyde and CO_2, and acetaldehyde forth reduces into ethanol. As *Z. mobiles* generate one ATP molecule for every glucose molecule, they divide in a very slow growth rate than *S. cerevisiae*, hence, resulting in a less reserves that are dissipated by development of undesirable biomass. Secondly, *Z. mobilis*exhibits less ATP concentration, so there is very low product constraint in ED pathway, as a result overall flow in pathway is enhanced [12]. *Z. mobilis* is also demanding due to its ability to generate ethanol much supremely than yeast and can resist some stressful conditions in a better way, together with high concentration of bioethanol. Such excellent distinctions enable *Z. mobilis* to transform hexose-sugar into bioethanol as a by-product in a rapid rate than yeast with about 97 % of ethanol yield. Another commercially beneficial bacterium involved in ethanol production is genetically modified *E. coli,* because of its capability to utilize a broad range of sugars and is a conventional commercial tool. Bacterium *E. coli* is an idealistic contender for analysis and research with metabolic engineering aspect, because it was comprehensively analyzed as a model microorganism. In late 1990s, there was seen wide range of study and research for determining a strain of *E. coli* that can be optimized for the development of bioethanol, but as it is very sensitive to atmospheric parameters such as temperature, pH and concentration of ethanol, it gets impended from being employed in the fermentation process [13]. The biogas technology is another biochemical conversion process, where biogas is synthesized through anaerobic breakdown of organic matter including crop residues, industrial and municipal wastes by the action of microorganisms, thereby resulting in liberation of CH_4, CO_2, and other gases in trace [15]. The nature of microorganism related to biogas production relies upon the type of substrate utilized and various atmospheric parameters. The sequence of simultaneous steps of the process that takes place during production of biogas demands actions of different microorganisms for different steps. Microorganisms involved in the process comprises of hydrolytic enzyme producing microorganisms, fermentative microorganisms or acidogens, acetate-producers or acetogens, H_2-producers, and methanogenic microbes (methanogens), where these all microbes acts synchronously leading to the production of biogas as a final product. It has been also observed that the amount of biogas produced increases from the initial to the middle of the digestion owing to increase the microbial count in this step, but there is decline in the production of biogas as soon as the process of digestion reaches its terminal phase because ultimately the microbial count also reduces which are responsible for biogas production [16]. There are several other anaerobic bacteria that have been studied well to exhibit the capability to decompose or commercially consume the lignocellulosic feedstocks as sole carbon sources for the biogas production includes genera *Clostridium, Ruminococcus, Fibrobacter, Acetivibrio, Proteobacteria, Actinobacteria, Spirochaeta, Thermotoga, Flavobacterium, Pyrococcus, and Streptomyces* [17].

3. SUBSTRATES UTILIZED BY MICROORGANISMS FOR PRODUCTION OF BIOFUELS

Generally, bioethanol production solely consumes carbohydrate or starch-based constituents that are derived from crop fields, spending most of its bioenenrgy enclosed within plant's lignocellulosic biomass. Unwanted sections of plants such as stems, leaves and timber woods comprises entirely of lignocellulosic substances and are very tough for microorganisms for degradation and if such lignocellulosic substances could be utilized, then they would be considered as a impactful bioenergy resource for the production of biofuels [18]. Many different forms of lignocellulosic feedstocks have been studied with action of defined microorganisms on them for certain parameters that have been optimized for attaining proper growth conditions of their particular microorganisms, enlisted in the Table 13.2. Generally, many forms of feedstocks can be utilized as a substrate for biogas production, if their major constituents are sugars, polypeptides and fats (Table 13.3). Preferred substrates must attain some specific features like being appropriate for fermentation, having a very high nutritive value, and bioreactors must be exempted of pathogenic microbes and heavy metals for field applications like pesticides, fertile manure, etc. There exist in wide range of residual wastes that forth can be utilized as resources or substrates, which are then supplied into the reactor to generate biogas. Majority of biodegradable natural components could be converted into biogas through anaerobic digestion, where the crude substances are usually residual feedstocks that comprises of solid wastes liberated from municipalities, sewage and solid wastes released from industries, residue from food processing, animal and crop manure, sewage from sludge digesters, and microalgae [24].

Table 13.2: Different Lignocellulosic Feedstock's with Associated Microorganisms for Bioethanol Production

Lignocellulosic Feedstock	Responsible Microorganism	Amount of Ethanol Produced (g/l)	References
Rice straw	*Aspergillusniger* and *S. cerevisae*	31.9	[19]
Pomegranate peel	*S. cerevisae* and *Pichiastipitis*	5.6	[20]
Banana stem	*Aspergillusniger, Trichodermareesei, Zymomonasmobilis*	3.5	[21]
Mango pulp	*S. cerevisae*	5.8	[22]
Wheat straw	*S. cerevisae*	23.85	[23]

Table 13.3: Different Categories of Residual Wastes Derived from Various Sources Along with their Associated Yield of Biogas

Category of Waste	Source	Yielded Biogas/ Biomethane (m^3/kg VS)	References
Agricultural Waste	Grass	0.190-0.198	[25]
	Wheat straw	0.270-0.288	[26]
	Cauliflower stalks	0.331	[27]
	Citrus waste	0.137	[28]
Municipal Waste	Activated sludge	0.376	[29]
	Sewage sludge	0.395	[30]
	Cattle manure	0.443	[31]
	Wastewater	0.232	[32]
Domestic Residues	Food waste	0.440-0.480	[33]
	Kitchen waste	0.700	[34]
	Orange peel	0.332	[35]
	Banana peel	0.277	[27]
	Onion skin	0.400	

Usually, the substrate utilized for its production might be categorized into three major groups comprising of agricultural residues, waste from municipalities and industrial wastewater and solid wastes.

4. BIOSYNTHESIS OF BIOETHANOL AND BIOGAS

Bioethanol synthesized from lignocellulosic feedstocks renders specific atmospheric, commercial and strategically profits [36]. As bioethanol is a vital alternative for fossil fuels, its production and demand is accelerating with upcoming days. Currently, ethanol acquires the leading commercial market because of its application as biochemical substrate or as a biofuel stabilizer or simply as a primary fuel [37]. Bioconversion of lignocellulosic feedstocks into bioethanol by-product mainly comprises of three stages (Fig.13.1).

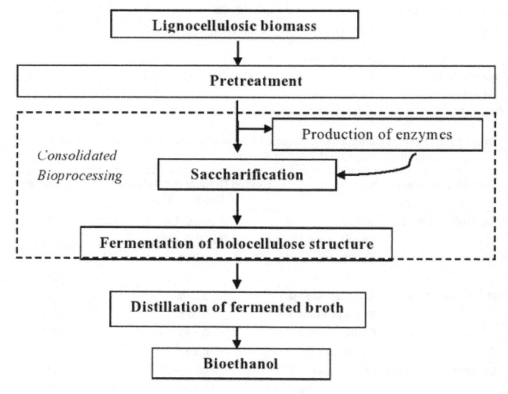

Figure 13.1: Process of Bioethanol Production from Lignocellulosic Biomass

In first stage, biomass pretreatment is done, so that cellulose and hemicellulose can be accessible for hydrolysis by degrading the complex structure of lignocellulose [38]. Second stage targets the hydrolysis of hemicellulose structure into fermentable sugars with the help of lignocellulosic enzymes i.e., cellulases, hemicellulases, etc. After detoxification step, third stage includes bioconversion of fermentable sugars into bioethanol [39].

For the production of biogas, a consortium of bacteria acts upon the waste biomass and degrade them under anaerobic conditions, as a result two major products are produced i.e., CH_4 and CO_2 (Fig.13.2). Anaerobic Digestion (AD) is a natural decomposition process of organic matter in anaerobic environment, in which production of biogas as an energy transporter and digested substances are generated. During this biochemical anaerobic digestion process, all steps are interrelated to one another. This process of anaerobic digestion can be grouped

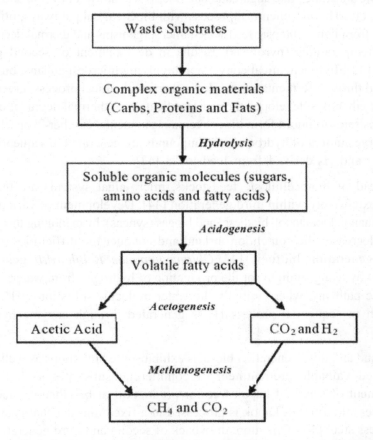

Figure 13.2: Production of Biogas from Waste Biomasses through Biochemical Conversion Process

into three major stages denoted as hydrolysis, acidification and methanogenesis that occurs and works in a synchronous manner within the same bioreactor and the vital and useful biogas production takes place in terminal methanogenic stage [40]. The composition of CH_4 in biogas lies between 50-75 % whereas the CO_2 constituent in between 25-50 % with certain deviations depending on the nature of substrate because of its chemical structure and biodegradability. Other remnants are H_2O, H_2S and NH_3. After purification of biogas, approximately 96-98% purity of CH_4 is attained, exhibiting relevant features to natural gas and can replace it as bio CNG [41].

5. SECOND GENERATION BIOFUELS

Second generation biofuels are usually produced from lignocellulosic biomass and non-edible substrates like sugarcane bagasse, wheat straw, domestic and municipal residues, etc. Thermo-chemical process with biochemical pathways, utilization of enzymes from lignocellulosic feedstocks, microorganisms and thermal decomposition at high temperatures (pyrolysis), results in development of second generation biofuels [42]. In biochemical conversion involving microorganisms, bioethanol is generated through fermentation while, in thermo-chemical process, green diesel or synthetic ethanol is developed through gasification or thermal degradation process. Such substrates to liquid form bioconversion pursue the Fischer-Tropsch pathway, where large amount of hydrocarbon compounds are generated in sequential process from CO and H_2 derived from feedstocks [43].

Instead of lignocellulosic feedstocks, micro-algal systems are utilized and studied extensively within third generation [44]. Development of various biofuels like bioethanol, biobutanol, biodiesel and biogas (syngas), by exploiting this microalgal system decreases the contention in land and conquer the difficulties of first and second generation biofuels [42].*Chlamydomonas Reinhardtii* generates bio hydrogen by using sunlight for the production of hydrogen from water, in aerobic-anaerobic pathway, where initially the water molecule splits into $2H^+$ and ½ O_2 and forth in sequential process H_2 is generated through enzyme hydrogenase [45].

Second and third generation biofuels exhibit some limitations as well, like need of defined valuable pretreatment of commercial substrates such as alkaline pretreatment of fibers of kapok tree for production of bioethanol effectively. In this process the fiber of kapok is processed in a fixed digester that is synthesized of stainless steel [46]. One more drawback of second and third generation biofuel comprises of entire cost productivity in total that varies based on the time and nature of substrate, therefore, restricting its viable benefit above conventional oils.

6. RELATED CHALLENGES AND POSSIBLE SOLUTIONS

There are different types of pretreatment methods that exhibit their own benefits and drawbacks; hence, there is no defined pretreatment method for all forms of biomass. It depicts that yet there is a gap between novelistic research in lab scale and practical implementation of these pretreatment techniques, so it is necessary to determine the efficacy of different pretreatment techniques, various structure of substrates, structural arrangement and linkage between the lignocellulosic biomass constitutions with their compatible pretreatment method. Furthermore, there is also not sufficient research available to enhance insight of the impact of every pretreatment on a structural and molecular basis. It is also to be noted that the study of interactions with their consequent hydrolysis and fermentation phase for various lignocellulosic feedstocks requires much more investigation in this field to modify the lignocellulosic structure essentially. To determine valid pretreatment method, it is essential to study the physical and chemical reaction techniques in much detailed manner; however, latest methods of pretreatment cannot be applied in pilot scale yet. Also the efficacy of biological or natural pretreatments as a commercial technique for enhancing the digestion is yet not clear because of the complex structure of microorganisms. In biological pretreatment, it is noted that the cost of the potential enzyme is reliably high, which is a major hurdle in implementation of pretreatment methods in a larger industrial basis. Also, it realize a need for more research to be carried to innovate new techniques and process in order to decrease the entire cost of the enzyme-based pretreatment and apparently to build it commercially compatible for large scale potential.

In this consideration, subsequent methods appear to be upcoming research drift in order to attain impactful and cost-effective use of enzymatic pretreatment for increasing production of biomethane from lignocellulosic feedstocks. Comprehending with the optimum activity of enzymes for ideal pretreatment of every form of lignocellulosic feedstock is a vital parameter to investigate in near future. Recycling of the expensive enzymes can be a second advancement method to decrease the concentration of used enzyme while pretreatment, which simultaneously reduces the operational cost up to much extent for production of biofuels [47]. Active enzyme recovery from the hydrolytic phase for another batch usage, thereby declining the comprehensive capital of enzyme utilized in hydrolytic step. As per analysis, it was noted that when the insoluble fractionated substrate was subjected to recycling process following enzyme-based pretreatment at beginning of the process, the enzyme utilized in process decreased by value of 30% with no notable alterations in resultant sugar yield [48]. Immobilization of enzymes can also be employed as a flourishing method in order to improve the recovery rate of catalytic activity during recycling of enzymes, therefore, immobilized enzymes have attained major consideration because of their optimistic features consisting of recycling and profitability [49]. Third approach can be advancement or establishment and implementation of genetically modified organisms (GMOs) to overcome the issues that come across from natural wild-strains. GMOs attain a novelistic ability to generate increased amount of enzymes, extremozymes that can resist harsh atmospheric/operational situations. Search for highly tolerant yeast and fungal microbes to such situations can be idealistic for accelerating the production of biomethane gas and for improved biofuel production through enzyme-based pretreatment that can be a massive commercial technique in near future [50].

REFERENCES

1. International Energy Agency, https://www.iea.org/reports/world-energy-outlook-2019; 2020.

2. Iodice P, Senatore A. Atmospheric pollution from point and diffuse sources in a National Interest Priority Site located in Italy. Energy Environ. 2016; 27: 586-96.https://doi.Org/ 10.1177%2F0958305X16665536.

3. Hussain A, Arif SM, Aslam M. Emerging renewable and sustainable energy technologies: State of the art. Renew Sust Energ Rev. 2017; 71: 12-28. https://doi.Org /10.1016/j.rser.2016.12.033.

4. Yusuf AA, Inambao FL. Progress in alcohol-gasoline blends and their effects on the performance and emissions in SI engines under different operating conditions. Int J Ambient Energy. 2018; 13: 1-7. https://doi.org/10.1080/01430750.2018.1531261.

5. Sheehan J, Camobreco V, Duffield J, Graboski M, Shapouri H. An overview of biodiesel and petroleum diesel life cycles. National Renewable Energy Lab. 1998; https://doi.org/10.2172/1218368.

6. Awasthi P, Shrivastava S, Kharkwal AC, Varma A. Biofuel from agricultural waste: a review. Int J Curr Microbiol App Sci. 2015; 4: 470-77.http://dx.doi.org/10.21741 /9781644900116-8.

7. Vasudevan P, Sharma S, Kumar A. Liquid fuel from biomass: an overview. J Sci Ind. 2005; 64: 822-831.

8. Lundgren A, Hjertberg T. Ethylene from renewable resources. Surf Renew Resour. 2010; 64: 109-126. https://doi.org/10.1036/1097-8542.BR0120141.

9. Ji M, Miao Y, Chen JY, You Y, Liu F, Xu L. Growth characteristics of freeze-tolerant baker's yeast Saccharomyces cerevisiae AFY in aerobic batch culture. Springer Plus. 2016; 5: 503.https://doi.org/10.1186/s40064-016-2151-3.

10. Mansouri A, Rihani R, Laoufi AN, Özkan M. Production of bioethanol from a mixture of agricultural feedstocks: biofuels characterization. Fuel. 2016; 185: 612–21.

11. Lin Y, Tanaka S. Ethanol fermentation from biomass resources: current state and prospects. Appl Microbiol Biot. 2006; 69(6): 627-42.https://doi.org/10.1007/s00253-005-0229-x.

12. Bai FW, Anderson WA, Moo Young M. Ethanol fermentation technologies from sugar and starch feedstocks. Biotechnol Adv. 2008; 26(1): 89-105. https://doi.org /10.1016 /j. biotechadv.2007.09.002.

13. Clomburg JM, Gonzalez R. Biofuel production in Escherichia coli: the role of metabolic engineering and synthetic biology. App Micro boil Biotechnol. 2010; 86: 419-34. https://doi.org/10.1007/s00253-010-2446-1.

14. Wang X, He Q, Yang Y, Wang J, Haning K, Hu Y, Wu B, He M, Zhang Y, Bao J, Contreras LM. Advances and prospects in metabolic engineering of Zymomonasmobilis. Metab Eng. 2018; 50: 57-73. https://doi.org/10.1016/j.ymben.2018.04.001.

15. Badiyya HM. Comparative study of biogas production from sugarcane bagasse and cow dung. Univers J Microbiol Res. 2018; 3(2): 127-131.

16. Aliyu S. Biogas production using different wastes: a review. Univers J Microbiol Res. 2019; 4(2): 75-82.

17. Azman S, Khadem AF, Lier JB, Zeeman G, Plugge CM. Presence and role of anaerobic hydrolytic microbes in conversion of lignocellulosic biomass for biogas production. Crit Rev Environ Sci Technol. 2015; 45(23): 2523-64. https://doi.org /10.1080 /10643389 .2015.1053727.

18. Bokinsky G, Peralta Yahya PP, George A, Holmes BM, Steen EJ, Dietrich J, Lee TS, TullmanErcek, D, Voigt CA, Simmons BA, Keasling JD. Synthesis of three advanced biofuels from ionic liquid-pretreated switch grass using engineered Escherichia coli. PNAS. 2011; 108(50): 19949-54.https://doi.org/10.1073/pnas.1106958108.

19. Yang P, Zhang H, Cao L, Zheng Z, Mu D, Jiang S, Cheng J. Combining sestc engineered A. niger with sestc engineered S. cerevisiae to produce rice straw ethanol via step-by-step and in situ saccharification and fermentation. Biotech. 2018; 8(1): 12. https://doi.org /10.1007/s13205-017-1021-1.

20. Demiray E, Karatay SE, Dönmez G. Evaluation of pomegranate peel in ethanol production by Saccharomyces cerevisiae and Pichiastipitis. Energy. 2018; 159:988–94.https://doi.org/10.1016/j.energy.2018.06.200.

21. Mustofa A. Bioethanol production from banana stem by using simultaneous saccharification and fermentation (SSF). In: IOP conference series: Materials Science and Engineering, IOP Publishing. 2018; 358(1): 012004. https://

ui.adsabs.harvard.edu /link gateway/2018MS&E..358a2004K/doi:10.1088/1757-899X/358/1/012004.

22. Barbosa CD, Lacerda IC, de Souza Oliveira E. Potential evaluation of Saccharomyces cerevisiaestrains from alcoholic fermentation of mango pulp. Afr J Biotechnol 2018; 17(28): 880-84. https://doi.org/10.5897/AJB2015.14847.

23. Brandenburg J, Poppele I, Blomqvist J, Puke M, Pickova J, Sandgren M, Rapoport A, Vedernikovs N, Passoth V. Bioethanol and lipid production from the enzymatic hydrolysate of wheat straw after furfural extraction. Appl Microbiol Biotechnol 2018; 102(14): 6269–77. https://doi.org/10.1007/s00253-018-9081-7.

24. Zabed HM, Akter S, Yun J, Zhang G, Awad FN, Qi X, et al. Recent advances in biological pretreatment of microalgae and lignocellulosic biomass for biofuel production. Renew Sust Energy Rev. 2019; 105: 105-128.https://doi.org/10.1016/j.rser.2019.01.048.

25. Rodriguez C, Alaswad A, Benyounis KY, Olabi AG. Pretreatment techniques used in biogas production from grass. Renew Sustain Energy Rev 2017; 68: 1193–1204.https://doi. org/10.1016/j.rser.2016.02.022.

26. Solé-Bundó M, Eskicioglu C, Garfí M, Carrère H, Ferrer I. Anaerobic co-digestion of microalgal biomass and wheat straw with and without thermo-alkaline pretreatment. Bioresour Technol. 2017; 237:89–98. https://doi.org/10.1016/j. biortech.2017.03.151.

27. Ji C, Kong CX, Mei ZL, Li J. A review of the anaerobic digestion of fruit and vegetable waste. Appl Biochem Biotechnol. 2017; 183(3):906–922. https://doi.org/10.1007/s12010-017-2472-x.

28. Patinvoh RJ, Lundin M, Taherzadeh MJ, Horváth IS. Dry anaerobic co-digestion of citrus wastes with keratin and lignocellulosic wastes: batch and continuous processes. Waste Biomass Valoriz. 2018; https://doi.org/10.1007/s12649-018-0447-y.

29. Ara E, Sartaj M, Kennedy K. Enhanced biogas production by anaerobic co-digestion from a trinary mix substrate over a binary mix substrate. Waste Manag Res. 2015; 33(6): 578–87. https ://doi.org/10.1177/0734242X15584844.

30. Silvestre G, Bonmatí A, Fernández B. Optimisation of sewage sludge anaerobic digestion through co-digestion with OFMSW: effect of collection system and particle size. Waste Manag 2015; 43:137–43. https://doi.org/10.1016/jwasman.2015.06.029.

31. Tsapekos P, Kougias PG, Kuthiala S, Angelidaki I. Codigestion and model simulations of source separated municipal organic waste with cattle manure under batch and continuously stirred tank reactors. Energy Convers Manag. 2018; 159: 1–6. https://doi.org/10.1016/jenconman.2018.01.002.

32. Guven H, Akca MS, Iren E, Keles F, Ozturk I, Altinbas M. Co-digestion performance of organic fraction of municipal solid waste with leachate: preliminary studies. Waste Manag. 2018; 71: 775–84. https://doi.org/10.1016/j.wasman.2017.04.039.

33. Zhang C, Su H, Baeyens J, Tan T. Reviewing the anaerobic digestion of food waste for biogas production. Renew Sustain Energy Rev 2014; 38: 383–392. https://doi.org /10.1016/j.rser.2014.05.038.\.

34. Jin Y, Li Y, Li J. Influence of thermal pretreatment on physical and chemical properties of kitchen waste and the efficiency of anaerobic digestion. J Environ Manag. 2016; 180: 291-300. https://doi.org/10.1016/j.jenvman.2016.05.047.

35. Siles JA, Vargas F, Gutiérrez MC, Chica AF, Martín MA. Integral valorisation of waste orange peel using combustion, biomethanisation and co-composting technologies. Bioresour Technol. 2016; 211: 173–182. https ://doi.org/10.1016/j.biortech.2016.03.056.

36. Chandel AK, Singh OV, Venkateswar Rao L, Chandrasekhar G, Lakshmi Narasu M. Bioconversion of Novel Substrate Saccharumspontaneum, a Weedy Material, into Ethanol by Pichiastipitis NCIM3498. Bioresour Technol. 2011; 102: 1709-14. http:// dx. doi.org/10.1016/j.biortech.2010.08.016.

37. Kumar R, Singh S, Singh O. Bioconversion of lignocellulosic biomass: biochemical and molecular perspectives. J Ind Microbiol Biotechnol 2008; 35: 377-91. http://dx.doi.org /10.1007/s10295-008-0327-8.

38. Singh A, Bajar S, Bishnoi NR. Enzymatic hydrolysis of microwave alkali pretreated rice husk for ethanol production by Saccharomyces cerevisiae, Scheffersomycesstipitis and their co-culture. Fuel 2014; 116: 699-702. http://dx.doi.org /10.1016 /j.fuel.2013. 08.072.

39. Gil N, Ferreira S, Amaral M.E, Domingues FC, Duarte AP. The influence of dilute acid pretreatment conditions on the enzymatic saccharification of erica spp. for bioethanol production. Ind Crop Prod 2010; 32: 29-35. http://dx.doi.org/ 10.1016 /j.indcrop.2010. 02.013.

40. Goswami R, Chattopadhyay P, Shome A, Banerjee SN, Chakraborty AK, Mathew AK, Chaudhury S. An overview of physico-chemical mechanisms of biogas production by microbial communities: a step towards sustainable waste management. Biotech. 2016; 6(1): 72. https://doi.org/10.1007/s13205-016-0395-9.

41. Solartetoro JC, Chacónpérez Y, Cardonaalzate CA. Evaluation of biogas and syngas as energy vectors for heat and power generation using lignocellulosic biomass as raw material. Electron J Biotechnol. 2018; 33: 52–62. https://doi.org /10.1016 /j.ejbt. 2018. 03.005.

42. Sims R, Mabee W, Saddler J, Taylor M. An overview of second generation biofuel technologies. Bioresour Technol. 2010;101(6):1570–80. https://doi.org/10.1016/j.biortech.2009.11.046.

43. Snehesh A, Mukunda H, Mahapatra S, Dasappa S. Fischer-Tropsch route for the conversion of biomass to liquid fuels-Technical and economic analysis. Energy. 2017; 130: 182–191.https://doi.org/10.1016/j.energy.2017.04.101.

44. Schenk P, Thomas-Hall S, Stephens E, Marx U, Mussgnug J, Posten C, Kruse O, Hankamer B. Second generation biofuels: high-efficiency microalgae for biodiesel production. Bioenergy Res. 2008; 1(1): 20–43.https://doi.org/10.1007/s12155-008-90088.

45. Melis A, Zhang L, Forestier M, Ghirardi M, Seibert M Sustained photobiological hydrogen gas production upon reversible inactivation of oxygen evolution in the green alga Chlamydomonasreinhardtii. Plant Physiol. 2000; 122(1):127–136. https://doi.org /10 .1104/pp.122.1.127.

46. Tye Y, Lee K, Wan Abdullah W, Leh C. Potential of Ceibapentandra (L.) Gaertn. (kapok fiber) as a resource for second generation bioethanol: effect of various simple pretreatment methods on sugar production. Bioresour Technol. 2012; 116: 536–39.https://doi.org/10.1016/j.biortech.2012.04.025.

47. Xing W, Xu G, Dong J, Han R, Ni Y. Novel dihydrogen-bonding deep eutectic solvents: pretreatment of rice straw for butanol fermentation featuring enzyme recycling and high solvent yield. Chem Eng J. 2018; 333: 712–72. https://doi.org/10.1016/j.cej.2017.09.176.

48. Weiss N, Borjesson J, Pedersen LS, Meyer AS. Enzymatic lignocellulose hydrolysis: improved cellulase productivity by insoluble solids recycling. Biotechnol Biofuels. 2013; 6: 1-14. https://doi.org/10.1186/1754-6834-6-5.

49. Agrawal R, Satlewal A, Mathur A.S, Gupta R.P, Raj T, Kumar R, Tuli D.K, 2018. Kinetic and enzyme recycling studies of immobilized â-glucosidase for lignocellulosic biomass hydrolysis. Agrawal Environ Eng Manag J. 2018; 1385-98. https://doi.org/10. 1126/science.1257859.

50. Lam FH, Ghaderi A, Fink GR, Stephanopoulos G. Engineering alcohol tolerance in yeast. Sci. 2014; 346:71-5.https://doi.org/10.1126/science.1257859.

CHAPTER - 14

BAMBOO AS A BUILDING MATERIAL FOR CLIMATE CHANGE MITIGATION

Vishal Puri

*Department of Civil Engineering
JC Bose University of Science and Technology YMCA,
Faridabad-121006, Haryana, India
Corresponding Author*

Construction Industry is one of the most polluting industry in the world. With the increasing population and industrialization, the demand and prices of conventional building materials such as steel and cement have risen exponentially. Their production also leads to dangerous levels of greenhouse gases generated in the atmosphere. Thus, there is a great need to look out for alternative sustainable building materials helpful for climate change mitigation. In this book chapter, author discusses the potential aspects of bamboo in construction industry for different structural elements along with the different advantages and limitations associated with it. Traditionally, bamboo is used as a sustainable construction material in many Indian states. Bamboo based walls with soil plaster is one such example. Many developing countries have also promulgated the application of bamboo in construction. Different researchers have showcased the potential of bamboo as a steel replacement material. Bamboo reinforced beams, bamboo reinforced columns, bamboo reinforced wall panels etc are few such structural elements wherein positive performance results have been obtained. Bamboo as a potential construction material can be hugely advantageous for climate change mitigation. Every ton production of bamboo absorbs more than a ton of carbon dioxide from the atmosphere making it a great carbon absorber. Along with its cost effectiveness and eco-friendly aspects bamboo has also high tensile strength comparable to mild steel. This chapter presents the different application aspects of bamboo in construction industry along with a comparative analysis with traditional building techniques. Based on the detailed analysis, comparative carbon credit analysis is also discussed to showcase the bamboo potential.

1. INTRODUCTION

Mankind is unable to provide housing to all even after several decades of industrialization, predominantly due to ever increasing population. More than 200,000 people a day are added to world population [1]. In India, more than 110 million houses are required by the year 2022 to achieve the dream of housing for all as per the estimates by Klynveld Peat Marwick Goerdeler report [2].To achieve the aim of housing for all there is a tremendous dependency on conventional building materials such as steel and cement. This has led to the spiral increase in their prices over the decades. Further, these materials are also observed to degrade the environment significantly. Construction Industry is one of the major contributors of greenhouse gases in the environment. Steel production is reported to produce more than two tons of carbon dioxide in the atmosphere. Similarly, cement is reported to degrade the environment on the same scale [3]. This has led to significant increase in greenhouse gases causing global warming and rising the average global temperature. Average temperature worldwide has already reached alarming levels causing the glaciers to melt and thus has increased the sea level. Due to which various coastal regions are at the verge of drowning.

Countries such as Maldives have already lost a large chunk of their land due to global warming [4]. To reduce such climate mitigation, International community has put different efforts. In the UN climate change conference (Paris Agreement, 2015), it was decided to persue efforts to limit the temperature increase even further to 1.5°C. This requires a lot of reduction in carbon dioxide emission into the atmosphere. This can be achieved by developing sustainable infrastructures. Sustainable infrastructures can be developed using sustainable building materials and sustainable construction techniques. Sustainable building materials such as bamboo fly ashes etc. are predominantly being used in different forms. These materials are highly energy efficient and cost effective and provide a potential alternative to conventional materials such as steel and cement.

2. BAMBOO IN CONSTRUCTION

Bamboo is a unique grass that grows naturally in many regions across the world (Fig.14.1). It grows in the tropical zones which mainly coincide with the developing countries. These countries usually have the highest rate of urbanization and population growth rate and thus the application potential of bamboo is immense [5]. Bamboo is reported to have more than 1250 species across the world [6] with some bamboo species growing at more than 91 cm per day [7]. These bamboo species are reported to attain a maximum height of 15 – 30 m.

Figure 14.1: Raw Bamboo

These species take about 3 – 8 years to reach its maximum strength with a diameter of 5 – 15 cm [8]. Every ton production of bamboo is reported to produce oxygen in atmosphere while absorbing about a ton of CO_2 from the atmosphere. This property makes it as an excellent material for climate change mitigation [3]. For developing infrastructure, bamboo is reported to have excellent material properties once used with proper surface chemical treatment (Fig 14.2). Tensile strength of some of the bamboo species is observed to be much more than mild steel [3]. Bamboo is reported to be 50 times more energy efficient than steel in terms of energy required for producing them [9].

Figure 14.2: Surface Treated Bamboo Splints

Further, its excellent strength to weight ratio makes it a good alternative over the steel. Different researchers have identified different advantages and limitations associated with the bamboo which are discussed in Table 14.1. The primary advantage of bamboo is its tensile strength. High tensile strength with low weight makes it as an excellent steel replacement material.

Table 14.1: Advantages and Disadvantages of Bamboo in Construction [10]

Advantages	Disadvantages
1. Excellent tensile strength	1. Prone to attacks by fungus and termites
2. Better Energy Efficiency	2. Mechanical properties vary with species
3. Excellent strength to weight ratio	3. Presence of node which affects tensile strength
4. Fastest growing plant (some species)	4. High rate of water absorption and shrinkage
5. Excellent carbon sink	5. Long term durability issues
6. Large availability	6. Flammability

2.1 Bamboo as Reinforcement in Structural Elements

Performance of bamboo as a reinforcement is evaluated in different structural elements by different researchers. Earliest research on bamboo started in 1914 wherein chow tested bamboo splits as reinforcement material for concrete applications [11]. Later several researchers researched upon Bamboo reinforced beams, Bamboo reinforced columns, bamboo arches and bamboo reinforced wall panels etc. Bamboo reinforced beams were developed by Ghavami in 1995 wherein he introspected the prospects of bamboo as reinforcement replacing the steel reinforcement [12]. He observed that bamboo reinforcement increased the ultimate load carrying capacity of the unreinforced concrete beam by 400%. The performance of bamboo reinforced columns was studied by Aggarwal et al [13]. They observed that bamboo reinforced column with 8% bamboo reinforcement provides the same strength and behavior under axial loading as compared to RCC column with 0.89% steel reinforcement. Author has showcased that bamboo bonding (Fig. 14.3) is dependent on different cement mortar constituents.

Author has also evaluated the performance of bamboo reinforced prefabricated wall panels. It was observed that addition of bamboo mesh as reinforcement in wall panels enhances the ductility in walls along with significant reduction in dead load and cost [14-15]. Similarly, different other researchers have showcased positive outcome of bamboo reinforcement in concrete [16-17].

2.2 Bamboo for Climate Change Mitigation

Bamboo as reinforcing building material has tremendous potential for climate change mitigation. Bamboo as a steel replacement can immensely save the environment from large amount of greenhouse gases released during steel production. A comparative effect of application of different building materials on environment is discussed in Table 14.2. It can be observed from the table that even if 30 percent replacement levels of steel with bamboo is achieved; a humongous 1121 Million tons of carbon dioxide emission can be saved from deteriorating the environment.

Figure 14.3: Bamboo Reinforced Cylinders for Pull out Testing

3. CODAL PROVISIONS

Different researchers have showcased positive potential of bamboo in construction industry. However, for large scale implementation major obstacle is the limited availability of codal guidelines. As the properties vary from bamboo species to species there are lack of standardization guidelines. Bamboo guidelines at the international level are still at nascent stage of development. Different bamboo resourceful countries have now taken a lead in defining the guidelines for bamboo application in construction though a very limited number of codal guidelines are available. Table 14.3 discusses the current codal guidelines provisions.

4. CONCLUSIONS

In this study the detailed prospects of bamboo as a building material were discussed. Bamboo when replaced with steel has immense potential for climate change mitigation along with cost and energy benefits. Even with a targeted 30 per cent reduction

Table 14.2: Comparative Energy Analysis of Different Building Materials

Building Material	Production	Energy Involved	Environmental Effect
Cement	4193 million metric tons worldwide [18]	Every ton of manufacture produces over two tons of CO_2 due to burning of fossil fuels. [3]	8386 million tons of CO_2 is released in atmosphere per year.
Steel	1869 Million Tonnes worldwide [19]	Every ton of manufacture produces over two tons of CO_2 due to burning of fossil fuels. [3]	3738 Million Tons of CO_2 is released in atmosphere per year.
Bricks	1.23 trillion bricks worldwide	Produces 800,000,000 t of CO_2 each year for 1.23 trillion bricks [20]	800 million tons of CO_2 each year [20]
Bamboo	20-million-hectare land under cultivation[21], average 1000 poles per hectare [22]	Each year, a hectare of Moso bamboo absorbs 5.1 ton of CO_2 [23]	102 million tons of CO_2 is absorbed yearly. Further utilization in construction industry will lead to an increase in Bamboo production and reduction in CO_2.
Fly ash	170 million tons in India [24]	Waste material produced by burning of coal. Produced in large quantities by thermal power plants.	70000 acres of land is presently occupied by ash ponds in India Itself.

Table 14.3: Different Codal Guidelines for Bamboo

Guideline	Provision	Country
• IS 6874 • Section 3B - National Building Code of India, • IS 15912	These guidelines provides the strength limits along with the bamboo joints and connection details, brief details about structural design using bamboo, details for testing of culm of bamboo to determine its physical and mechanical properties	India
JG/T 199: Testing method for physical and mechanical properties of bamboo used in building (PRC MoC, 2007)	This standard provides testing method for physical and mechanical properties of bamboo. This standard focuses on the culm wall for mechanical tests instead of full culm.	China
• NTC 5407, • NTC 5525	These guidelines provide details on structural joints with *Guadua Angustifolia Kunth*. It is also provides the methods to test the properties of locally available bamboo.	Colombia
ASTM D5456: 2013	It recognizes the laminated veneer bamboo as a structural product and also provides the guidelines on manufacturing standards and test methods.	USA
INEN 42	It discusses the processing, selection, construction and maintenance. However, it does not include any design guidance details.	Ecuador
• ISO 22156: Bamboo – structural design • ISO 22157: Bamboo – Determination of physical and mechanical properties	The basic design using full culm construction was provided in ISO 22156: This guideline is further supported by ISO 22157-1: Bamboo – determination of physical and mechanical properties – part 1: requirements (ISO, 2004b), providing the test methods, and ISO 22157-2 (ISO, 2004c), which is a manual for determining material properties.	International

replacement a humongous 1121 million tons carbon dioxide production can be saved. However, for massive application tremendous efforts are required to develop standard guidelines along with further research leading to standardized application.

REFERENCES

1. World Population Balance. World Population Statistics, http://www.worldpopulation balance.org/faq;2020 [accessed 14 October 2020].
2. KPMG Insights. Decoding housing for all by 2022, https://home.kpmg/in/en /home /insights/2014/09/decodinghousing.html;2014 [accessed 14 October 2020].
3. Bhalla S, Gupta S, Sudhakar P, Suresh R. Bamboo as green alternative to concrete and steel for modern structures. J. Environ. Res. Dev. 2008; 3(2).
4. The Hindu. A sinking Feeling, https://frontline.thehindu.com/the-nation/article 30188693.ece; 2009 [accessed 14 October 2020].
5. Javadian A, Wielopolski M, Smith IF, Hebel DE. Bond-behavior study of newly developed bamboo-composite reinforcement in concrete. Construction and Building Materials. 2016; 122: 110-7.
6. Scurlock JM, Dayton DC, Hames B. Bamboo: an overlooked biomass resource?. Biomass and Bioenergy. 2000; 19(4): 229-44.
7. Guiness World Records. Guinness book of world record, http://www.guinness worldrecords.com/world-records/fastest-growing-plant;1999 [accessed 14 Oct. 2020].
8. Li Y, Shen H, Shan W, Han T. Flexural behavior of lightweight bamboo–steel composite slabs. Thin-Walled Structures. 2012; 53: 83-90.
9. Fang HY, Ghavami K. Low-cost and Energy Saving Construction Materials. In: Yan Xiao, Masafumi Inoue, Shyam K. Paudel, editors. Modern Bamboo Structures, London: CRC Press; 1984.
10. Puri V. Development of Prefabricated Bamboo Reinforced Fly Ash Replaced Green Mortar Wall Panels. In: Doctoral dissertation, IIT Patna; 2019, India.
11. Chow HK. Bamboo as a Material for Reinforcing Concrete. In: Doctoral dissertation, Massachusetts Institute of Technology; 1914, USA.
12. Ghavami K. Ultimate load behaviour of bamboo-reinforced lightweight concrete beams. Cement and Concrete Composites. 1995; 17(4): 281-8.
13. Agarwal A, Nanda B, Maity D. Experimental investigation on chemically treated bamboo reinforced concrete beams and columns. Construction and Building Materials. 2014; 71: 610-7.

14. Puri V, Chakrabortty P, Anand S. Flexural behaviour of bamboo-reinforced wall panels with varying fly ash content. Magazine of Concrete Research. 2020; 72(9): 434-46.
15. Puri V, Chakrabortty P, Anand S, Majumdar S. Bamboo reinforced prefabricated wall panels for low cost housing. Journal of Building Engineering. 2017; 9:52-9.
16. Korde C, West R, Gupta A, Puttagunta S. Laterally restrained bamboo concrete composite arch under uniformly distributed loading. Journal of Structural Engineering. 2015; 141(3): B4014005.
17. Kankam JK, Ben-George M, Perry SH. Bamboo-reinforced concrete beams subjected to third-point loading. Structural Journal. 1988; 85(1): 61-7.
18. Statista. Major countries in worldwide cement production, https://www.statista.com/ statistics/267364/world-cement-production-by-country/; 2020[accessed 14 Oct. 2020].
19. World steel association. Global crude steel output, https://www.worldsteel.org/media-centre/press-releases/2020/Global-crude-steel-output-increases-by-3.4—in-2019.html; 2020 [accessed 14 October 2020].
20. Fast company. Printable bricks could cut World's carbon emissions, https://www. fastcompany.com/90185191/printable-brick-could-cut-worlds-carbon-emissions-by-at -least-800-million-tons-a-year-update; 2010 [accessed 14 October 2020].
21. Sharma S. Bamboo Industry of North East India. In: Employment News. vol. XXXIX No.17, Ministry of Information and Broadcasting; 2015.
22. Business Diary. Bamboo Production Guide, http://businessdiary.com.ph/4409/bamboo-production-guide; 2018 [accessed 14 October 2020].
23. Inesad Development Roast. What bamboo can do about CO_2, http://inesad.edu.bo/ developmentroast/2013/05/what-can-bamboo-do-about-co2; 2013 [accessed 14 Oct. 2020].
24. Mission Energy. Fly Ash Utilization, https://missionenergy.org/flyash2013/whyflyash.html; 2013 [accessed 14 October 2020].

INDEX

A

Acetogenesis 51, 52, 136
Acidification 51
Acidogenesis 51, 64, 136
Acidogenic 18
Agro residue 231, 232, 234
Agro-waste 231, 232
Alcohols 51, 121, 168,
Algal biomass 131, 132, 137, 139, 140, 143, 144, 168
Alkalinity 19, 137
Anaerobic 17, 19, 26, 32, 71, 73, 131, 146, 147, 148, 256, 261, 266
Anaerobic Digestion 136, 137, 261
Application of biochar 173

B

Bagasse 10, 11, 78, 199, 231, 232, 233, 234, 238, 262
Bamboo bonding 272
Bamboo guidelines 273
Bamboo in construction 269, 270, 272, 273
Bamboo reinforcement 272
Bamboo Species 270, 272, 273,
Binders 248
Biochar 173, 174, 175, 176, 177, 178, 179, 180, 183, 184, 185, 186, 187
Biochar production technologies 175
Bio-CNG 18, 70
Biodiesel 3, 14, 15, 142, 143, 143, 144, 154, 170, 172
Bioenergy 1, 14, 113, 131
Bioethanol 139, 140, 141, 142, 150, 253, 255, 257, 259, 260, 261, 263, 265, 266, 267
Bio-fertilizers 35, 45
Biofuel 254, 255, 256, 260, 262, 264, 268
Biogas 6, 7, 8, 9, 15, 15, 17, 18, 18, 19, 21, 21, 22, 22, 23, 24, 25, 26, 27, 28, 29, 30, 31, 32, 33, 34, 35, 36, 36, 37, 38, 39, 39, 40, 41, 43, 45, 46, 47, 49, 50, 51, 53, 54, 55, 57, 58, 58, 59, 61, 62, 63, 65, 67, 68, 69, 70, 71, 73, 73, 74, 75, 136, 137, 146, 148, 255, 259, 261, 265

Biogas plants 40, 45
Biogas production 15, 32, 36, 46, 61, 265
Biogas refrigeration 32
Biogas slurry 35, 36, 38, 39, 40, 45
Biogas slurry processing 40, 45
Biogas technology 17, 18, 35, 36
Biomass 2, 11, 50, 77, 78, 82, 85, 88, 89, 99, 102, 113, 115, 117, 131, 232,
Biomass densification 232, 248, 249
Bio-Oil 10, 14, 174, 175, 189
Biopolymer 133
Biorefineries 140
Blend 3, 139
Bligh and dyer method 143
Bricks 23, 246, 274
Bulk density 105, 124, 231, 235, 238, 241, 242, 249
Bureau of Indian Standard 201, 203

C

C/N ratio 20, 20, 21, 137, 138
Calorific Value 123, 245
Carbon capture 154, 169
Carbon dioxide 66, 89, 90
Carbon dioxide production 276
Carbon monoxide 89, 90
Carbon sequestration 173, 183, 191
Carbon sink 2, 6, 272
Carbonization 175, 240
Cascades 160
Cement 23, 27, 274, 276

Characteristics of microalgae 133
Chlorella 138, 142, 146, 147, 148, 155, 171
Clean cooking 193, 201, 204, 206
Climate 1, 2, 28, 32, 36, 49, 49, 50, 102, 188, 201, 207, 210, 222, 225, 226, 269, 270, 271, 272, 273, 275, 277
Climate mitigation 270
Closed systems 153, 157, 158, 160, 163
Co-digestion 137, 138, 139
Cogeneration 2, 11, 12, 13, 31
Conditions 119, 120, 182
Construction industry 78, 269, 270, 273, 274
Controlled cooking test 214
Cookstove 193, 194, 197, 199, 202
Cookstove classification 194, 195
Crop residue 80, 104, 107, 168, 169, 174, 187, 201, 205
Cultivation 133, 134, 135, 144, 155, 159, 161, 162, 164, 165
Cultivation of algal biomass 133
Cultivation of microalgae 133, 134, 146

D

Deenbandhu 22, 23, 27, 38, 77
Design 22, 27, 146, 151, 164, 206, 207, 238, 251
Devolatilization 115, 166, 118, 120, 121, 128
Dissemination 49, 194, 202, 210, 211, 215, 227
Distillation 140

E

Economics 45, 59, 77, 247
Economics of biochar 187
Efficiency 54, 106, 212, 213, 222, 245, 272
Emissions 50, 188
Energy 1, 8, 13, 84, 104, 206, 215, 245, 251, 274, 276
Energy plantation 3, 6
Entner-Duodoroff pathway 256, 257
Exergy efficiency 199
Exothermic 87, 114, 116, 117
Extremozymes 264

F

Fermentation 140, 141
Field 209, 211, 212, 220, 228, 235
Fixed Dome 22, 27
Flexi model 25, 27
Floating dome 23, 27
Flowability 123
Fluidized bed 64, 90, 91, 93, 97
Fly Ash 274, 276, 277
Folch method 143
Forced draft 195, 196, 198, 199, 202, 203, 222
Fossil fuel 2, 30, 49, 50, 131, 132, 184, 253
Fossil fuels 14, 32, 33, 50, 77, 78, 102, 153, 154, 231, 232, 253, 254, 255, 260, 274, 274
Fractionation 143, 144

Fuel 5, 29, 30, 31, 104, 105, 213, 214, 245
Furnace 4

G

Gasification efficiency 106
Gasifier 10, 90, 91, 92, 94
Genetically modified organisms 264
Global energy 231
Glycolytic pathway 256
Gobar gas 18
Greenhouse gas 1, 14, 32, 49, 158, 185, 188
Greenhouse gases 3, 32, 67, 232, 254, 255, 269, 270, 272

H

Hemicellulose 116, 120
Household air pollution 193, 195, 199, 205
Housing 270, 276, 277
Hydraulic retention time 21, 33, 56, 137, 145
Hydrogen 10, 60, 89
Hydrolysis 50, 51, 136, 141, 142

I

Improved biomass 193, 203

K

Kitchen performance test 214
KVIC 8, 22, 23, 24, 27, 38

L

Laboratory Tests 211
Light intensity 132, 154, 156, 162, 163
Lignocellulose 53, 63, 71, 255, 261, 268
Lignocellulosic feedstocks 148, 258, 260, 262, 263
Liquid fraction 36, 42, 45
LPG 8, 15, 101, 195, 204, 221

M

Macroalgae 132, 133, 135, 136, 137, 142, 144
Mesophilic 20, 38, 53, 54
Methanogen 52, 64, 71, 136
Methanogenesis 52, 64, 136
Micro algae 153, 154, 156
Microalgae 132, 133, 137, 142, 154, 159
Microorganisms 63, 256, 258
Microwave 62, 126
MNRE 21, 23, 27, 29, 36, 38, 202
Moving bed 126
Multi-Fuel 201
Municipal solid waste 17, 18, 78, 79, 98
Municipal waste 6, 105, 132, 154, 158, 259

N

National biomass cookstove initiative 203
National program for improved cookstove 202
Natural draft 196, 199, 201, 202, 222
NNBOMP 8, 21, 27

O

Open pond 153, 158, 162, 163
Organic fertilizer 17, 18, 33, 35, 36, 39, 40, 82
Organic loading rate 21, 137

P

Pelletization 45, 169, 123, 233, 235, 249
Performance evaluation 42, 209, 210, 212, 216, 225
Petrodiesel 3
pH 19, 38, 51, 53, 137, 154, 156, 162, 181, 256
Photo bioreactor 135, 160, 161,
Physicochemical properties of biochar 180, 181
Piston press 236, 239
Plasma 95, 98
Policy 5, 194, 195, 202, 203, 205
Power 102, 212
Pretreatment 62, 136, 139, 143
Producer Gas 10, 77, 83, 89, 101, 103, 106, 199, 202,
Protocol 211, 215, 216, 222, 223, 224, 225, 228, 229
Proximate analysis 60, 243, 249
Pyrolysis 82, 83, 85, 93, 130, 180, 188, 189, 190

R

Raceway pond 134, 161, 163, 166
Reactor 26, 86, 99

Renewable energy 1, 2, 8, 27, 35, 36, 136, 169, 175, 202, 253
Residence time 92, 97, 105, 114, 115, 117, 126, 129, 175, 176, 179, 181, 196
Roller press 237, 238, 240
Rotary 126, 176, 179, 235, 238
Rotary drum reactor 126

S

Screw conveyor 126
Screw press 36, 40, 42, 44, 46, 233, 237, 239, 241, 247, 249
Separate hydrolysis and fermentation 140, 141
Simultaneous saccharification and fermentation 140, 141, 265
Size reduction 19, 62, 234
Soil conditioner 183, 187
Solid Biomass 10, 77, 121, 193, 194, 201, 204, 213, 215
Solid fraction 36, 42, 44, 45
Specific gasification rate 106, 199
Steel 270, 274
Super capacitors 185
Sustainable infrastructure 270

T

Technology 28, 31, 32, 40, 126, 214
Tensile strength 271
Testing 129, 211, 214, 224
Thermal conversion 69, 82
Thermal efficiency 58, 92, 104, 193, 196, 199, 201, 204, 213, 222, 224, 245
Thermoelectric generator 196, 206
Thermophilic 20, 153, 137
Torrefaction 113, 115, 117, 120, 122, 125, 127
Transesterification 142, 143, 144,
Triglycerides 144

U

Unnat Chulha Abhiyan 203, 205

W

Waste material 17, 174, 231, 232, 274
Waste water treatment 26, 154, 158, 168, 175, 181, 184, 184, 187
Water boiling test 209, 214, 215, 222,
Wood 19, 33, 36, 78, 83, 90, 102, 104, 105, 114, 124, 125, 195, 201, 210, 213, 216, 221, 223, 234